**WITHDRAWN**
UTSA LIBRARIES

# REVENUE PROTECTION:
*Combating Utility Theft & Fraud*

# REVENUE PROTECTION:
## *Combating Utility Theft & Fraud*

Karl A. Seger, PhD

Copyright ©2005 by PennWell Corporation
1421 South Sheridan Road
Tulsa, OK 74112-6600 USA
**800-752-9764**
**+1-918-831-9421**

sales@pennwell.com
www.pennwellbooks.com
www.pennwell.com

Director: Mary McGee
Managing Editor: Steve Hill
Production / Operations Manager: Traci Huntsman
Production Manager: Robin Remaley
Assistant Editor: Amethyst Hensley
Book Designer: Clark Bell

Library of Congress Cataloging-in-Publication Data Available on Request
Seger, Karl A.
Revenue Protection: Combating Utility Theft and Fraud
ISBN 1-59370-039-3

All rights reserved. No part of this book may be reproduced, stored in a retrieval system, or transcribed in any form or by any means, electronic or mechanical, including photocopying and recording without the prior written permission of the publishers.

Printed in the United States of America
1 2 3 4 5  09 08 07 06 05

# Table of Contents

1. **Revenue Protection: An Overview of the Challenge** .......... 1
   - Case Studies .......... 1
   - Revenue Protection From an Economic Perspective .......... 6
   - Revenue Protection as a Risk Management Issue .......... 7
   - Categories of Theft and Fraud .......... 8
   - Organized Crime and Revenue Protection .......... 12
   - Summary .......... 14
   - Recommended Actions .......... 15
   - References .......... 16

2. **Organizing For Revenue Protection** .......... 17
   - Revenue Protection Models .......... 17
   - Responsibilities, Departments, and Coordination .......... 21
   - Proactive Activities .......... 24
   - Reactive Activities .......... 29
   - Developing Policies .......... 33
   - Summary .......... 33
   - Recommended Actions .......... 34

3. **Crime Scene Investigation — Theft of Service** .......... 35
   - Detecting Theft of Service .......... 35
   - Principles of Investigation .......... 39
   - Rules of Evidence .......... 41
   - Processing the Crime Scene .......... 45
   - Documenting the Crime Scene .......... 49
   - Summary .......... 52
   - Recommended Actions .......... 52
   - References .......... 53

4. **The Fraud Investigation** .......... 55
   - The Fraud Problem .......... 55
   - Categories of Customer Fraud .......... 59
     - Residential fraud problems .......... 59
     - Commercial fraud problems .......... 59

> Combating Utility Customer Fraud ............................................. 62
> > Application for service ............................................. 62
> > Bad checks ............................................................. 62
> > Planned bankruptcy ................................................. 64
> > Out of business—back in business ........................... 65
> > Skips ....................................................................... 66
> Getting Below the Tip of the Iceberg ............................... 67
> Summary .......................................................................... 68
> Recommended Actions ..................................................... 68
> References ........................................................................ 69

## 5 The Paper Chase and Revenue Recovery .................................................. 71

> Sources of Information .................................................... 71
> > Government sources ................................................. 72
> > Business sources ....................................................... 73
> > Legal and safety sources ........................................... 73
> > The Internet as an investigative tool ....................... 78
> Computing the Back Bill .................................................. 80
> Collecting Lost Revenues ................................................. 86
> Summary .......................................................................... 88
> Recommended Actions ..................................................... 89

## 6 The Investigative Interview .................................................................. 91

> Planning for the Interview ............................................... 91
> The Interview .................................................................. 97
> > Field interviews ........................................................ 97
> > Office interviews ...................................................... 98
> Detecting Deception ......................................................... 102
> > Pronouns ................................................................. 106
> > Nouns ...................................................................... 106
> > Verbs ....................................................................... 106
> > Repeated accounts ................................................... 107
> Advanced Interview Techniques ....................................... 107
> Summary .......................................................................... 110
> Recommended Actions ..................................................... 110
> References ........................................................................ 111

**Table of Contents**

**7  Investigations at Potentially Dangerous Accounts..................................................113**
   The Problem..................................................................113
   Drug Manufacturers .....................................................115
   Indoor Marijuana Growing .........................................120
   Defusing Anger and Aggression .................................123
   Summary.......................................................................126
   Recommended Actions.................................................127
   References ....................................................................128

**8  Investigative Challenges and Tools ...........................................................................129**
   Internal Investigations .................................................129
   Criminal Intelligence ...................................................132
   The Scenario ................................................................135
   The Challenge and the Tools........................................138
   Summary.......................................................................146
   Recommended Actions.................................................    147
   References ....................................................................147

**9  Working with Law Enforcement and Going To Court............................................149**
   Establishing Liaisons....................................................149
   Solvability Factors .......................................................153
   Major Crime Scene Investigations ..............................155
   Preparing for and Going to Trial .................................158
   Summary.......................................................................132
   Recommended Actions.................................................163
   References ....................................................................163

**10  Evolving Challenges..................................................................................................165**
   The Impact of Deregulation ........................................165
   Changes In Metering ...................................................167
   Sarbanes-Oxley Act of 2002 .......................................171
   Protecting Investigators and Investigations ..............172
   Summary.......................................................................175
   Recommended Actions.................................................176
   References ....................................................................177

Appendix A Position Description .................................................................................... 177

Appendix B Procedures to Handle Utility Theft and Fraud Situations Found in the Field
 by Service Personnel ................................................................................. 181

Appendix C Draft Ordinance ........................................................................................ 189

Appendix D Twenty-fouth Guam Legislature .................................................................. 193

Appendix E Revenue Protection Sample Web Page ......................................................... 197

Appendix F Revenue Protection Investigation Report ....................................................... 199

# Dedication

To Suzanne, my wife and partner for the past 40 years—the person who keeps the home fires burning while I travel the world chasing utility thieves and other reprehensible characters.

I also wish to express my sincere appreciation to the editors and staff at PennWell Corporation. I am grateful for their patience and gentle prodding as deadlines came and went, and for their input and professionalism at all stages in the writing and publishing of two books.

**Revenue Protection:** Combating Utility Theft & Fraud

# Introduction

Energy theft in the United States costs consumers billions of dollars each year. This reflects billions of dollars that are built into their utilities' rates bases and paid for by honest consumers. In other countries, energy theft may result in losses that exceed 50% of the energy generated. In a world where energy is a critical commodity, the need to combat energy theft should be a goal and even an obligation for every utility.

As a Chief Financial Officer for a utility commented, developing a revenue recovery program is like finding change that has been lost in the pillows of a couch. It is money the utility let slip out of its pockets and that it needs to recover.

The emphasis of this book is on developing a structured revenue protection program where the goals are to identify and recover revenues lost to energy thieves and customer fraud. An effective program will also help to identify revenues lost to metering problems and billing errors. These losses almost always exceed losses resulting from theft and fraud.

Larger utilities should have a full-time staff dedicated to revenue protection and recovery. Smaller utilities will assign these duties to personnel in other positions who put on their revenue protection investigator hats when a potential theft is discovered. In both cases, the primary objective of the program is to recover lost revenues and to convince first offenders not to attempt to steal utility services in the future.

An effective revenue protection program will decrease losses resulting from theft and fraud and help the utility recover revenues already lost. It will add dollars to the utility's revenue stream and help to deliver services to consumers at a reasonable price.

# Revenue Protection: An Overview of the Challenge

## Case Studies

His method was simple. He attached a jumper to his overhead electrical service, taped it on, and a portion of the load bypassed the meter. It was simple and it worked. He started stealing electricity during World War II for the three apartments and the small wood shop that he owned. He wasn't caught until February 2001 when the now 91-year-old customer called to complain about an outage. The utility found the jumpers when responding to the complaint. The statute of limitations limits the utility to charging the customer with only seven years of stealing and he has been back billed $82,000. If he could be billed for the entire time, at today's rates his bill would be $655,999. It's bad enough that the customer was stealing for at least 56 years, but it's even worse that the utility never detected the potential theft during that period.

In the 1990s a utility on the east coast of the United States decided to randomly check the meters at grocery stores. They were looking for metering problems but also found meter tampering. At the Primo Grocery Store they found the owner began stealing electricity three months after he opened the store. Before he was caught he had stolen $30,000 worth of services. In a plea bargain the utility agreed to bill him $20,000 for the lost services. Similarly, at the Fairway Market the utility found the owner had stolen more than $140,000 in electric services. He faced a sentence of up to 10 years in prison and a $100,000 fine.

Police in Florida arrested a man they say created a device that hundreds of utility customers used to tamper with their electric meters and reduce their power bills. According to police reports the landscaper who worked on the side as a *fixer* had 200 customers who paid him $25 per month to activate his device, which allowed approximately half of the homeowner's electricity to bypass the meter. The average client had saved roughly $12,000 off their bills in the past four and a half years. These customers are being billed for the energy used but not paid for. The fixer will most likely go to prison.

A fixer is someone who rigs meters for a living. It may be a local electrician with an additional income or it may be someone who does this full time. Full-time fixers usually work on commercial and industrial accounts. The part-time fixer often works on residential accounts.

*Meter tampering* refers to adjustments made directly to the meter to reduce consumption. *Diversion* refers to methods used to bypass the meter. For our purposes we will focus on the discovery and investigation of tampering and diversion as they apply to energy theft; however, the same approaches are used in combating other forms of utility theft.

Utility theft costs utilities and their honest customers. An electric and gas theft in New York resulted in an estimated $165,272 in electricity and $20,984 in illicit gas usage. That utility estimates that it is losing nearly $60 million dollars a year to utility thieves. This means that nearly $6 of every customer's bill is lost to energy theft. Many utilities estimate that between 0.5% and 3.5% of their gross revenues are lost to utility thieves. However, in many emerging nations where energy resources are critical these losses may be 20% of gross revenues or more.

In Kenya the government passed the Electric Power Act of 1997, which provides that a conviction for stealing power is punishable by fines of up to SH30,000 (US$380) and/or up to two years in prison. An initial inspection of meters by the Kenya Power and Lighting Company resulted in the arrest of 675 suspects at residential, commercial, and agricultural sites.

In South Africa in a squatter camp outside of Durban, parents caution their children to wear shoes and to look out for electric wires on the ground where people are stealing energy. In one brief period three children were electrocuted before authorities raided the camp. When authorities arrived they found that a number of people without authorized electrical connections had electric appliances. Their illegal connection had obviously been removed just before the raid. Other South African thieves are more sophisticated. In one area they were connecting directly to the local substation, running a line to their homes for free power.

In Israel power to the Knesset and all the traffic and street lights in the area were disrupted for several hours as a result of illegal connections to the power system. There was a 1997 power outage in a major section of Rehov Ruppin as a result of power theft at a camp for the homeless when authorities were unable to correctly project and respond to peak load demands.

In 1997 the Twenty-Fourth Legislature of the territory of Guam passed emergency legislation to address the power theft problem. Section 1 of the statute states that, "The legislature finds that, according to the Guam Power Authority [GPA], there is an increasing number of unauthorized electrical connections or 'illegal hookups' on GPA power lines. Theft of electrical current is a growing problem, and it is at a point where something has to be done." The revised Public Law #34-31 provides a reward of $500 to any person who reports an illegal power hookup resulting in the collection of fines and penalties or a conviction. The name of the person reporting the crime remains confidential. Penalties for energy theft range from $2500 up to a maximum of $25,000.

Australian utilities estimate they are losing $15 million to utilities thieves each year, much of it to indoor marijuana growers. Anyone caught stealing energy in that country can be fined up to $20,000.

In New Zealand energy fines are less steep, but thieves may also go to jail. A man found guilty of stealing in Christchurch was sentenced to one month's periodic detention and ordered to pay $168.57 in reparation. An energy thief in Oxford was sentenced to three month's periodic detention and had to pay $2,816.08 to the utility.

In India, a country with a serious energy problem, it is estimated that at least 21% of the electricity generated is stolen. Unaccountable losses in Delhi are even higher, rising from 23.6% in 1993 to 47.9% in 1999.

To combat these losses the country passed antitheft legislation in 2000. Penalties in the new law include the following:

- mandatory punishment of three months to five years
- penalty of US$105 to $1060
- disbarred from the energy supply for up to two years
- collusion of utility staff is explicitly recognized as an offense

The law provides for special tribunals in each district headed by an additional district judge and an appellate court at the state level. It requires a summary trial and disposal of cases within six months.

One utility in India developed a unique approach to reduce theft in areas controlled by dadas, local organized crime. The dadas, or mafia men, bid for local power contracts and pay a lump sum for power supplied to a master meter, from which residential customers receive energy. Revenue collection is left to the local dada who receives a 25% commission. A utility in South America has arranged a similar agreement with local organized gangs who become the "power distributor" in some of the more dangerous areas. The gangs have their own penalties for individuals caught stealing energy from them.

In 2003 Shaheed Badruddin Sheikh, 32, was arrested for the third time in seven months for defrauding the Brihanmumbai Electric Supply and Transport Undertaking (BEST). The utility believes he has been stealing and illegally selling energy for at least five years. According to one of his customers, Sheikh charged slum dwellers Rs 200 to run a wire to their house and a monthly electricity fee of Rs 300. At least one of his hookups caused a fire that resulted in one person dead, two injured, and the destruction of 150 slum dwellings.

The Pakistani government activated 30,000 members of the armed forces to help fight power theft in 1998. According to the preamble of a new ordinance, "the menace of stealing electricity is rampant, necessitating calling on the armed forces of Pakistan to act in aid of civil power to suppress the said menace and to provide the speedy trial of offenses." It was estimated that 40% of the energy generated was stolen. The offenders were not all impoverished; of the 87 members of the upper house in Parliament, 49 were accused of energy theft for a cumulative annual loss of $232,000.

In China, the Northeastern Utility Group estimates that approximately half of their service industries are engaged in energy theft. Many connect themselves directly to distribution transformers. In 1998 when they launched a major campaign against utility thieves 9,769 offenders were identified.

A scrap dealer in Stanley, England, learned that he could save on his electricity bill by plugging into a street lamppost and powering his outbuildings, workshops, four color televisions and a number of other appliances at the utilities' expense. When asked what he was charged with, he was told, "About 50,000 volts." He was found guilty.

Indoor marijuana growers in England also like to save on their light bills. An indoor grower in London was caught routing the power for his growing operation around the meter. When police entered his flat they found marijuana with a street value of nearly $100,000. He was charged with possession of marijuana for trafficking and with the theft of electricity. Marijuana growers complain that their biggest expense other than equipment is electricity. Of course, that's only if they are paying for the energy they use.

In the Caribbean, Jamaica Public Service Company (JPSC) estimates that 8% to 10% of the energy generated is stolen. They conduct early morning blitzes into high crime areas, accompanied by the police, and arrest energy thieves. They also aggressively target commercial and industrial energy theft. In one month alone, the utility conducted investigations on 4,755 suspected cases and found 1,348 irregularities. Like many other utilities around the world, JPSC found that some of the best fixers on the Island are their own employees. Other Caribbean islands have found that establishing and maintaining an effective revenue protection program can significantly reduce loses. Barbados, for example, has maintained a revenue protection program for several decades and total line losses there are less than 6%.

In Latin America the problem is not just theft but refusal to pay electric bills. Consumers in one country developed the "I hate the light" Internet site to vent their frustrations. In some places vigilante groups intercept line crews and threaten them if they don't repair the transformers in their neighborhoods.

The Canadian Electricity Association issued a report on the Extent of Energy Diversion on Customer Premises for Canadian Utilities. The report is based on a study conducted across Canada. The study results indicate, "Deviations which will lead to diversion are definitely occurring across Canada. The average rate for these deviations is 1.36%." This accounts for an annual revenue loss of $130,475,320.

# Revenue Protection From an Economic Perspective

Across the United States approximately 80% of the revenue protection cases investigated are residential. The other 20% are industrial, commercial, and agricultural. However, 80% of the dollars lost to energy thieves is lost at industrial, commercial, and agriculture accounts. Because the utility must have a certain fiscal operating margin, when revenues that should be collected are lost to theft those revenues usually end up in the rate base. Honest customers then pay for what the few dishonest customers are stealing.

Arizona Public Service Company (APS) conducted one of the most recent studies on revenue losses. The goal of the study was to determine the dollar amount lost to theft and diversion. The professionally conducted study included a stratified random sampling of 550 meters from across the APS service area, including 64 commercial accounts. The data were collected by APS electric service employees who checked the voltage at the targeted accounts and made approximately 53 entries on a form designed for the study. The study was conducted between April and June 2000. The results found in the study indicated the following:

- Definite meter tampering: 0.72%
- Probable meter tampering: 1%
- Actual dollars lost: $330,148
- Actual loss revenue percentage: 0.0215%

- Probable loss dollars: $7,637,131

- Probable loss revenue percentage: 0.4965%

- Total actual/probable loss dollars: $7,967,279

- Total actual/probable loss revenue percentage: 0.5180%

Losses at commercial accounts, only 12% of the meters examined, accounted for the greatest loss. Of the $7.9 million actual/probable loss, $5.1 million was attributed to commercial accounts.

Another important finding in the results of the APS study is that 6.5% of the meters in the study had some type of maintenance problem. A revenue protection program will also help to identify metering and billing problems that are costing the utility more than theft and fraud combined. For example, the investigator at a utility in the southeast back bills approximately $250,000 a year for revenues lost to power thieves. When a new major municipal facility went online, however, the utility was using a multiplier of one on the billing and the error resulted in a loss of more than $300,000 per month before it was detected.

The utility's investment in a revenue protection program should return approximately two to three dollars for every dollar invested. Some utilities claim a seven to one return. According to the International Utility Revenue Protection Association (IURPA), its member utilities average one investigator for every 52,000 customers. As we will discuss in chapter 2, many smaller utilities with 25,000 or more meters find that assigning a full-time employee to revenue protection is a solid investment.

# Revenue Protection as a Risk Management Issue

A 45-year-old utility customer who had been disconnected for non-payment strapped on a pair of climbing spikes and scaled a wooden utility pole to reconnect his power. The transformer fuse had been removed, and although he successfully refused the transformer he made the mistake of closing the disconnect switch by hand. He was reportedly drinking with friends just before the incident. One of those friends said his friend was just trying to replace a street lamp-like safety light on the pole.

In another incident, a customer placed jumpers behind the meter to decrease his light bill. The jumpers shorted out, and his house was destroyed by fire. The fire also resulted in the death of his seven-year-old daughter.

In Canada, a man tinkering with the lights at his small indoor marijuana growing operation was electrocuted. He had rigged the electricity for his 1,000 watt grow lights to bypass the meter. He died for 24 marijuana plants. On the U.S. west coast an electrician was trying to steal power when he was electrocuted, set on fire, and killed. At the same utility two motor home owners died in one year while trying to steal power from street lights.

Revenue protection is a risk management issue. If the utility is not making a concerted effort to protect its employees and customers from electrocution and fire, will the utility be responsible if someone is killed or injured because of that neglect? How about this scenario: The utility knows for certain that a customer is stealing power. They let him continue for a month so they can compare the readings on his meter with the reading on a secret check meter they have installed for billing purposes. During the month the customer's illegal actions result in a fire in which he and his family are killed. The fire also destroys two neighboring homes. Does the utility bear any responsibility in this case?

In the case of our customer who was killed after re-fusing his transformer, the family tried to sue the utility for illegally disconnecting him and not providing adequate protection at the pole. Although the case never went to court, the utility's attorney had to respond to the initial complaint.

## Categories of Theft and Fraud

There are two types of utility thieves: honest thieves and dishonest thieves. *Honest thieves* may be an oxymoron and it is unique to utility services theft. The honest thief wouldn't think of shoplifting or any other petty crime. If a store clerk gave this person $5 too much in change, the honest thief would return it, then go home and turn the electricity meter upside down. Honest thieves believe that

- No one is looking.
- Nothing will happen if I'm caught.
- Everyone is doing it.
- It's not stealing; it's taking.

In many cases the honest thief is correct, no one is looking and nothing will happen if they are caught. Consider the case of two U.S. utilities, each of which has approximately 65,000 customers. Utility A has a full-time revenue protection program that is constantly identifying energy thieves and collecting lost revenues. Utility B, located adjacent to Utility A does not have a revenue protection program because no one in their service area is stealing. Of course, they never really looked.

Not everyone is stealing, but friends and relatives may brag that they know how to steal energy. Maybe they do. Maybe they don't. As for the theory that "it's not stealing; it's taking," here is a quote from an honest thief who explained, "Stealing is when you take something you don't need. But when you take something you do need, that's not stealing, that's taking."

Those who would categorize themselves as honest thieves include individuals who reconnect themselves after being disconnected for nonpayment and people who share how-to-steal information with friends and neighbors. This category also includes, from the customer's perspective, fraudulent customers who change the name on the account after being disconnected for nonpayment. *Dishonest thieves,* however,

- Steal everything they can
- Hang around with other criminals
- Have family members who are also criminals
- Expect to be caught occasionally

In one case a dishonest thief was caught stealing gas at his business. When confronted by the gas company, he paid the bill in full and started stealing electricity instead. If you find a dishonest customer stealing energy and they have natural gas, they will probably be stealing that as well as cable television services and probably a portion of their water. A good way to identify these thieves is to monitor the local newspaper. Who was arrested for what this week? Consider checking some of those meters, especially indoor marijuana grows and commercial accounts involved in other illegal activities.

There are four major categories of energy theft and fraud: residential, commercial, industrial, and agricultural.

Residential theft falls across the entire economic spectrum, from the needy to the greedy. In some cases the utility should assume some responsibility for residential theft. Take the case of the fixer who was rolling back the meters on at least 200 residential meters each month. The fact that he was able to do this undetected demonstrates that the utility didn't have a meter seal program. Neither the outside nor the inside of the meters was sealed. He was not even caught by the utility but reported to the police by someone he tried to solicit as a new customer. His accounts were each paying him $25 a month for which he would roll back between $10 and $15 worth of usage each month.

If you find one family member stealing energy, try to identify the locations of other family members on your system. If you find several people in the same neighborhood, you should inspect every service in the area. If you have identified one case of energy theft, you have just located the tip of the iceberg. There are other cases to be found.

Commercial theft occurs at high energy users, marginally profitable businesses, businesses experiencing economic problems, businesses engaged in other illegal activities, and businesses with greedy owners. If the owners of a business where energy theft is found own other businesses, inspect the meters at all locations. If they are stealing at one location, they are probably stealing at all of the locations.

Consider the franchised fast-food restaurant chain that brought a fixer to its annual owners meeting. Following the meeting theft was found at locations across North America. It should be emphasized that most of the franchise owners did not engage in energy theft. But those that decided to use what they had learned were stealing at all their franchises.

In another case the utility noted that the local police had raided eight bars owned by the same group of local criminals. The raid was a result of suspected illegal drug sales and prostitution. When the utility checked the electric meters at the eight locations the morning after the raids, they found that all eight bars were stealing energy.

If you find energy theft at a business, check the owner's home as well. And if you find energy theft at the home of a business owner, check the meters at the business. If they are stealing at one location, they are probably stealing at the other.

Theft at industrial locations can result in the loss of millions of dollars. Wires were crossed on the metering of a major industrial plant resulting in a significant reduction in the metered energy. The utility didn't uncover the situation until several years later and the loss was millions of dollars. The company claimed the utility made the mistake when the meters were installed and refused to pay the entire amount. When the case went to court, the company introduced into evidence the policy of the utility stated that if there was a metering problem the back billed or credit would be for 90 days of usage. The judge followed the utility's policy and awarded it 90 days of back billing, not the millions of dollars they asked for. The policy has since been changed.

A utility in the Midwest received a telephone call from a utility in the east. A large industrial user in the east was stealing utility services at all its plants. During the investigation they learned the company owned three plants in the Midwest and telephoned to suggest that the Midwestern utility check the services at those plants. Services had been tampered with at all three plants and the Midwestern utility recovered almost $100,000 in lost revenue.

Commercial and industrial meters should be read by metering personnel trained to inspect these services and detect problems. Most meter readers are not trained to detect problems at current transformer and potential transformer meter locations. Problems could result in substantial losses if they are not detected early.

Theft at agricultural locations can take several different forms. In some cases agricultural accounts sign up for reduced load management rates and then bypass the load management equipment. They enjoy the lower rate but don't have to worry about the equipment shutting down during peak loads. At one location the farmer ran part of his load off the wiring to the security light behind one of the out buildings. He paid the flat rate for the security light and didn't pay anything for the electricity used at the outbuilding or a nearby trailer.

# Organized Crime and Revenue Protection

The Federal Bureau of Investigation (FBI) defines organized crime as "any group having some manner of formal structure and whose primary objective is to obtain money through illegal activities." In other words, organized crime is when a group of criminals come together, select a leader, and engage in crime. Those crimes include utility theft. Organized crime groups commit energy theft at

- Legitimate businesses they own
- Illegal businesses they operate
- By providing fixer services to other businesses

Remember the eight bars we discussed that were all stealing energy? They were legitimate businesses (at least they had business licenses). The bars were owned by a local organized crime group. In another city a local organized crime family controlled all the vice-related crimes in a certain area. The police conducted a year-long investigation to build cases against the family members on a number of charges. During the investigation the criminals hired an undercover officer to roll back the consumption on their meters each month. When numerous charges were brought against them, those included theft of utility services.

Illegal business operated by organized crime present special dangers to utility employees. An employee might find a diversion unaware that the location is an indoor marijuana grow operated by a local crime group. This is a particular problem in certain parts of Canada where Asian crime groups have taken over this business. In central Canada police raided four locations and arrested eight Asian suspects charged with producing marijuana for the purposes of trafficking. They are considering charges under the antigang law, which makes it illegal to belong to a criminal organization. The electric meters (called *hydro meters* in Canada) had been bypassed at each location so that the electricity used to power the lighting, watering systems, and fans didn't register. The raids netted 1,427 marijuana plants, almost 13 kilos of packed pot ready for distribution and sale, and $100,000 in cash. Street value of the pot seized is more than $2 million. The street value of the electricity stolen is unknown.

Organized crime may use fixers to steal energy or other utility services at their legitimate and illegal businesses. They sometimes sell the fixer's services to other legitimate businesses. In some cases the fixers themselves are the criminal group.

A group of criminals operated for a brief time on the U.S. east coast adjusting the meters each month at a number of commercial and industrial accounts. They were paid a percentage of what they saved the account. The accounting books that were discovered when they were arrested listed almost 100 accounts in one state alone. Police and the utilities in that region may never know how many meters the criminals adjusted each month.

In western Canada where *B.C. Bud* is known as a high-quality, high-potency cannabis, it is often swapped for cocaine or methamphetamine (meth) with U.S. criminal gangs. The exchanges often take place in Montana. The meth that is traded pound for pound for B.C. Bud is worth $12,000 to $15,000 in the U.S. Drug dealers in California trade a pound of cocaine for a pound of B.C. Bud. Street value of the cocaine is between $6,000 and $9,000.

Methamphetamine labs are a concern around the world, especially in the United States and Canada. Although they have become a cottage industry, often operated by a couple of meth addicted operators, they represent a special danger to utility employees. When methamphetamine is produced, dangerous chemical are used, and an even more dangerous residue is produced. These residues are often carelessly discarded. A meter reader who comes in contact with the chemicals, either on clothing or skin, could suffer serious consequences but may not know it until later.

There is also a danger of an explosion and fire at a meth lab. If there is a short in the doorbell when a utility employee goes to the house, it could result in an explosion that would injure the utility employee and might result in the deaths of the occupants.

A third danger from methamphetamine labs is the people who operate them. As noted, many are addicted to the drug, which has a definite impact on the user's personality. Police refer to this as *Meth Rage Violence*. One officer reports that he has never seen a more devastating drug that leads to systemic violence and a total deterioration in the addict's quality of life. Police quickly learn that addicts will kill in a heartbeat if their meth is at stake.

Utility employees working in the field need to be aware of the dangers of marijuana grows and methamphetamine labs. Local police forces should be asked to present information on the problem in the area and employees should learn what to look for in the yard or trash that may indicate the presence of one of these operations. If they have a reason to be suspicious at any location, they should back off and let their supervisor know what they discovered. The police are then notified.

## Summary

Almost every utility is experiencing revenue protection problems. If the utility is not organizing to prevent losses and recover revenues, then it is passing the loss on to its honest customers. It is also ignoring customer practices that could result in fire, injury, and death.

If the utility is only addressing the problem at residential accounts, then it may not be identifying its major dollar losses. Remember the results of the APS study where $5.1 million of the $7.9 million was lost to commercial accounts even though those accounts only represent 12% of the meters in the study.

Each chapter concludes with a list of recommended actions to develop or review your revenue protection program. Use them as a guideline. In most utilities you will form a working group, review the recommended actions, and determine which of these are applicable to your situation.

## Recommended Actions

- Review your utility's policies regarding revenue protection utility, back billing, and related procedures. You should also review pertinent statutes and laws. These should be reviewed with an attorney. Policies will be further discussed in chapter 2 and a sample policy is found in Appendix B.

- If you have a revenue protection program in place, are you identifying commercial and industrial theft? If you are in a rural area, are you looking for agricultural theft? Residential theft is the easiest to find, but the real dollars are lost to commercial and industrial accounts.

- Review the cases you have already discovered and look for patterns. Have you found several cases using similar methods in the same neighborhood? If so, check all the meters in the neighborhood. Have you identified several members of the same family? Check all the family members you can identify. If you found a business stealing, did you check the meter at the owner's house? If you found the owner stealing at home, did you check the business?

- Who reads your commercial and industrial meters? If you are using meter readers, then consider forming a separate unit of meter personnel—in a small utility a meter person—to read those meters. You're not just looking for diversion or tampering, you're also looking for maintenance and other problems that are costing you money. Commercial and industrial accounts pay the utility's bills. Monitor them carefully.

- If you are the electric utility—and gas, cable, and water are separate—when you find an energy theft, suggest to the other utilities that they check their services at the location. Do not tell them you caught the customer stealing. Just suggest they check the services. Ask them to do the same for you.

- Consider inviting the local police to brief your field employees on the marijuana and methamphetamine problems in your area. Ask them to show these employees what to look for in the field that may be an indicator of these illegal operations. Let employees know who to contact when they find a suspicious situation or location.

# References

*Associated Press.* 2004. Police arrest man accused of creating device to steal energy. June 4. http://www.wftv.com/newsofthestrange/3381154/detail.html (accessed July 1, 2004).

Bhatia, B., and D. K. Prasad. 2003. Issues and challenges in controlling electric theft and leakages—a case study of Andhra Pradesh, India. *A Renewed Agenda for Energy, Energy Week 200,* The World Bank. http://www.worldbank.org/energy/week2003/papers.html (accessed June 1, 2004).

Claiborne, W. 2001. Energy thefts climb amid power crisis; Utilities pass along losses in higher rates. *The Washington Post.* May 26. http://www.highbeam.com/library/doc3.asp?DOCID=1P1:44776852&num=2&ctrlInfo=Ro (accessed July 1, 2004).

CNN. 2001. Utah man, 91, accused of stealing power for decades. *CNN.com.* February 15. http://www.cnn.com/2001/US/02/15/power.theft/index.html (accessed July 7, 2004).

Culwell, J. 2001. Research study quantifies energy theft losses. *Metering International*, Issue 1. http://www.metering.com/archieve/011/22_1.htm (accessed June 1, 2004).

Dillon, D. 2001. Press Release. July 18. http://www.nassauda.org/dawebpage/pressreleases/runpfrelease.htm (accessed July 1, 2004).

Krishnakumar. 2003. Man defrauds BEST of Rs 1 core. *Mid-Day Multimedia.* http://ww1.mid-day.com/news/city/2004/june/85650.htm (accessed July 1, 2004).

Federal Bureau of Investigation. FBI Investigative Programs Organized Crime. http://www.fbi.gov/hq/cid/orgcrime/glossary.htm (accessed July 3, 2004).

Owen, B. 2004. Pot farms tied to Asian ring. *Winnipeg Free Press.* March 5.

Sanderson, B. 1991. Man faces 10 years in power theft. *The Record* (Bergen County, NJ). February 7. http://www.highbeam.com/library/doc3.asp?DOCID=1P1:22575544&num=4&ctrInfo=R(accessed July 1, 2004).

Seger, K. 2002. Energy utilities around the world address this growing problem. *Metering International* Issue 1. http://www.metering.com/archive/021/20_1.htm (accessed July 1, 2004).

Van Doran, J. 1977. Plea deal in theft of energy. *The Record* (Bergen County, NJ). July 15. http://www.highbeam.com/library/doc3.asp?DODCID-1P1:22400945&num=2&ctr1Info=Ro (accessed July 1, 2004).

# 2

# Organizing for Revenue Protection

## Revenue Protection Models

There is no *correct* way to establish a revenue protection program at your utility. There are a number of different effective approaches. How you organize your program will depend on the size of your utility, the resources available and frankly, the politics within the utility. However you organize your program, you should make sure that you include each of the components shown in figure 2.1.

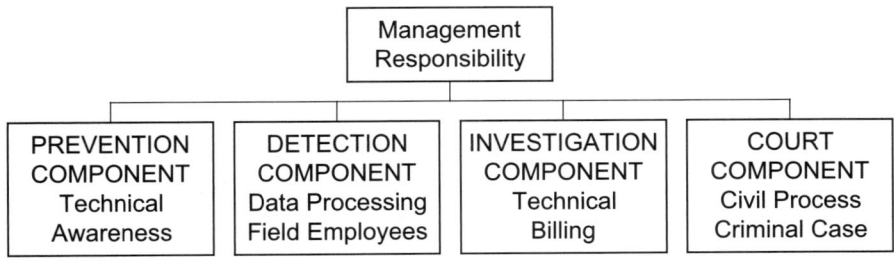

Fig. 2–1. Revenue protection program components

The most important aspect of the prevention component is a well-managed seal program. All your meters must be sealed, and seal integrity should be checked periodically—at least once a year. In addition, seals must be controlled. This can be accomplished by assigning seals to employees by serial number. If a seal is found out of place, such as on a tampered meter, the employee should be held accountable for how the seal got out of the employee's control. Used seals, including demand seals, should be brought back to the meter shop for disposal.

Meters must also be controlled. They shouldn't be left around the back of a truck or at services that have been inactive for some time. Meters that are damaged or taken out of service for any reason should also be disposed of at the meter shop. An astute meter shop foreman at a Caribbean utility keeps his used meters and seals in a locked area until he is ready to discard them. The meters are then damaged so that they are unusable, and both meters and seals are taken to the dump. He personally watches as the old meters and used seals are covered by the bulldozers at the dump.

The second element of the prevention component is awareness. Utility employees should be aware of revenue protection problems as well as the program goals. This includes office employees who may hear about an energy theft. It is even more important for field employees who must know how to detect a potential tampering or diversion. Some utilities include a presentation on revenue protection in the new employee orientation and the annual ethics briefing. Field employees should receive update briefings biannually. The update briefing alerts employees to the methods of diversion and tampering in the field and reminds them that every report results in a follow up investigation. It also gives the utility an opportunity to provide recognition to those employees who are expertly detecting revenue protection cases.

Customers should also be conscious of your revenue protection program, but they certainly don't need to learn methods for stealing utility services. Press releases and other customer communications should emphasize two basic points. First, if a neighbor is stealing energy, everyone else in the community is paying for it. Revenues lost to energy thieves result in higher rates for all the utility's customers. Second, energy theft is dangerous. The thief may create a situation that results in a fire, an injury, or death. It may not be the thief who is injured or killed but an innocent bystander. Provide a system where customers can report energy thieves anonymously.

If you are making the public aware of revenue protection, then utility employee training becomes even more important. Employees will be approached by customers claiming to have read or heard about someone stealing energy and asking how it's done. The employee should be trained to respond that they are not allowed to discuss methods used because they are dangerous, and if someone is stealing, the rest of the customers are paying for the loss. They can also explain to the customer that energy theft is a crime and people can go to jail if they are caught.

The key to the detection component is employee awareness and motivation. Field employees must know what to look for and how to report a suspicious situation. They should be alert for the three potential responses when a possible theft is detected:

1. If there is an indication of potential tampering but no obvious evidence, such as a cut seal or unusual marks around the meter or meter base, information on the location and potential indicators can be reported and inspected the following day.

2. If tampering is obvious, such as a meter upside down in the socket, it should be reported immediately via telephone or radio so that it can be investigated as soon as possible. The field employee will usually continue reading meters or working orders.

3. If the tampering has created a dangerous situation, such as an open meter base, the field employee should report the incident immediately and remain at the location until someone arrives to investigate and render the situation safe. If the customer approaches the field employee before the investigator or crew arrives, the field employee should explain that there is an unsafe situation and that a crew is on the way to correct it. Do not get into a confrontation with the customer.

Your computer department plays an important role in detecting potential energy theft. The department should print a monthly exceptions report listing accounts whose consumption is higher or lower than expected, and those accounts will be inspected to determine if there is a metering problem. There are a number of formulas used to identify the accounts on the report. Some utilities compare the current month's billing with the same month the previous year. Others compute the average for the last 12 months after removing the months of the highest and lowest usage. The report should include all commercial and industrial accounts and show the constant, or multiplier, used to compute the bills at those accounts. The report is reviewed by the meter department, revenue protection, and billing.

Full-time investigators will conduct the revenue protection investigation. Smaller utilities will assign investigative responsibilities to an employee in a different full-time position such as a meter technician. In either case, the investigator should be trained in crime scene investigation and evidence collection. If you are using a part-time investigator with another position, use the same person for all investigations. Use the same person in the billing department to compute the back bills on all revenue protection cases. This will help to develop an expertise in these cases in the event that this person has to testify in court. The same is true for the meter technician. If a meter technician other than the investigator tests meters in these cases, use the same individual for every investigation. Again, you will have established the expertise of the meter technician should you have to take a case to court.

Some municipal utilities have the police investigate all their revenue protection cases. If this is your policy, try to work with the same officer on all cases and assign the same meter technician to work with the officer. This will help develop an effective investigative team.

Will every case go to court? Some utilities take every case to court as a deterrent, whereas others prefer to avoid court when possible. Although every case may not go to court, you have to treat each case as though it will from the time you suspect tampering or diversion. This is discussed in chapter 3.

In criminal court, when a customer is accused of energy theft, you must be prepared to prove your case. In civil court you are not accusing any specific person of the theft, you are stating that it occurred. As a result, the utility did not bill or collect the revenues for all the energy used, and the person brought to court used the unbilled energy. Preparing cases for court is discussed in chapter 9.

Consider the following policy with regard to referring cases to criminal court. With first-time offenders, the objective is to recover revenues. If the customer can be presented with a bill for the lost revenues and it is paid, then court is avoided. If not, a claim is filed against the customer in civil court. Once the case is resolved, the customer's name should be in a suspense file and the service inspected again in 12 months.

Unfortunately some customers will become repeat offenders. Repeat offenders must have criminal charges filed against them. Fixers should also have charges, hopefully multiple charges, filed against them. Some of these cases will go to both criminal court, to punish the offender, and civil court to collect the lost revenues.

The final component in our model is *management responsibility*. Obviously someone has to be in charge of the total revenue protection function. Where the program is placed in your utility will depend on the structure of your organization.

# Responsibilities, Departments, and Coordination

The implementation of organization is the second component of the program. In some utilities, revenue protection is considered customer service or billing. In others it is a meter department function. If the utility has a security department, revenue protection responsibilities may be assigned there. In a few cases, revenue protection is deemed internal auditing. Where the program is placed is less important than the fact that most of the departments within the utility must work together for a successful program.

Let's consider the roles of the following departments:

- Metering
- Meter reading
- Customer service
- Billing and collections
- Security
- Customer information service—computer department
- Service workers and line crews

The meter department has the necessary expertise to conduct the technical aspects of the investigation. They can also identify technical methods used to combat tampering and diversion and approaches to securing your meters. They will be responsible for testing meters in energy theft cases to determine how much energy was not registered as a result of the tampering or to determine that the meter was running correctly in diversion cases. Meter technicians are especially important in cases involving commercial and industrial metering. Meter readers and service personnel are not usually trained to identify problems at those locations. Some utilities have meter technician crews that inspect all commercial and industrial meters at least once a year. Others assign meter technicians to read commercial and industrial meters monthly rather than assigning the responsibility to meter readers.

Even if they are not reading commercial and industrial meters, meter readers are your first line of defense in combating energy theft. They cover your entire service area monthly; so they need to know what to look for and how to report suspicious situations. Some utilities require meter readers to read inactive meters to verify that they are still inactive and that the meter is still there. Meter readers can also be required to inspect the integrity of the meter seals on a scheduled basis, at least once a year. They should also receive feedback when they report a potential or obvious case so that they know their reports are followed up, even if it is determined that it was not energy theft. Meter readers and other field personnel should receive revenue protection refresher training every six months.

Customer service personnel interact with customers daily on the telephone and in person. They should be aware of energy theft and report suspicions to their supervisor. Utility customer service personnel become very good at telling when a customer is "running a game" or lying to them. If a customer has a high bill complaint that is not resolved the way the customer thinks it should be, the account could be monitored for the next several months. The same action could be taken with a customer who not granted an extension on a bill. These are disgruntled customers and some of them will try to get even with the utility through energy theft.

The Billing and Collections Department plays a proactive and a reactive role in revenue protection. This department can help identify customers who have been on the monthly disconnect list for years who suddenly start paying their bills on time. Most likely their habits haven't changed, but there is a reason they don't want you to pull their meter. This department will help compute the back bills in revenue protection cases and collect the lost revenue.

If your utility has a security department, it will play a key role in the program. If revenue protection is under some other department in the utility, then security may not be involved in every investigation. They should still be involved in any investigation that is obviously going to court, repeat offenders and fixers, investigations where other crimes such as marijuana growing have been committed, and any investigations involving employees or contractors. Security personnel are experienced investigators, and they have an established liaison with the law enforcement agencies in your area.

The Customer Information Services (CISs), or computer department, is responsible for the monthly exceptions report. They can also assist in analyzing targeted categories of accounts. For example, you might target fast food restaurants where CIS could conduct an analysis of all the customers in this category to determine if there is an unusually low usage at any of the accounts. They can also establish whether there has been an unusual change in consumption that was undetected by the exceptions report.

Service personnel and line crews also have revenue protection responsibilities. Service personnel are working with meters every day and will identify many of your cases. Line personnel should know what to look for in the field and how to report a suspicious situation.

An effective revenue protection program requires the task force approach just discussed. Initially, representatives from each department should meet monthly to discuss progress made and needs that have been identified. Once the program is operational, the meetings can be held quarterly. People attending the meetings should be encouraged to suggest innovative approaches rather then just reviewing what has taken place. For example, if there is a downturn in certain aspects of the local economy, the utility could determine the category of business most affected by the downturn and monitor those accounts.

# Proactive Activities

Proactive activities, the second component of the program, are measures taken or decisions made as the utility is developing its revenue protection program. Reactive activities take place after a potential theft is detected. Your utility must decide, which of these are most appropriate for your situation.

## Develop a revenue protection program

Six months ago you caught the mayor's son stealing energy. You corrected the situation and told him not to do it again. A month ago you caught Joe Citizen stealing and have referred the case to criminal court. Is this malicious prosecution? After all, you are not treating everyone equally.

It's not malicious prosecution if you can demonstrate that six months ago you did not have a program in place and now you do, and show that you are treating everyone the same based on the policy and procedures of your program. This is referred to as a relief from past practices. You should have a policy that assigns revenue protection responsibilities within the utility and that outlines the procedure for your program. A sample policy and procedures is provided as an appendix.

## Amnesty period for employees

Establish the date that your revenue protection program officially begins. Then let employees know that they have until that date to report any problems with their meters or seals. The problems will be corrected, no questions asked. After that date any employee involved in energy theft, either at their service or operating as a fixer, will be suspended or terminated and referred for prosecution. A utility employee who is stealing utility services is stealing from their employer. As with any other form of employee theft, this is a major violation of the trust given the employee. In one large utility with a workforce of more than 10,000 employees, in excess of 250 employees requested new meters or seals during the amnesty program. When the program began, the first thing the revenue protection department did was to inspect the meters of all employees. Another 250 employees who continued to steal energy were identified.

## Amnesty period for customers

If an amnesty period is offered for employees, it should also be offered for customers. Customers are informed that the revenue protection program goes into effect on a specific date. If any customer has a meter or seal problem, they should let the utility know and the problem will be corrected. State clearly that after the amnesty period is over, energy thieves could be prosecuted and go to jail. Let them also know that if a neighbor is stealing, they are paying for it and that the theft could result in a dangerous situation.

When Guam enacted its energy theft act in 1998, the legislature allowed the utility 45 days to conduct a public awareness campaign before the theft act went into affect. An amnesty period of 90 days went into effect following the 45-day *educational period*. Any customer who requested assistance to correct a condition during the amnesty period had the corrections made without penalty. Every customer caught stealing after the amnesty program was arrested, booked at the jail, and subject to fines and other penalties.

One of the objectives of a country in Central America in developing its revenue protection program was to generate immediate revenues. They accomplished this through a unique customer amnesty program. Customers who turned themselves in during amnesty had the back bill computed for only the previous 12 months regardless of how long they had been stealing. They were then given 12 months to pay the bill. More than $5 million was back billed during this period. Since the end of the amnesty period, customers caught stealing have had their service immediately terminated. Service is not reconnected until the entire back bill is paid in full. There are no credit arrangements for utility thieves.

## Training for employees

Begin training with new employee orientations—just a brief presentation on the importance of revenue protection. For office and other inside employees, an annual update on revenue protection is appropriate. If your utility has an annual ethics briefing, the presentation could be made at that time.

Outside employees should receive a brief revenue protection presentation every six months. These could be conducted during safety meetings. Review what to look for and how to report suspicious situations. Also remind employees not to confront customers at these locations. Unless there is a dangerous situation, they should report what they find and continue with their work orders or reading meters.

## Public awareness campaign

Larger utilities often establish an energy theft hotline—a telephone number that a customer can call to anonymously report energy theft. A few utilities, like Guam Power Authority (GPA), offer rewards to customers who report thieves and the name of the person remains anonymous. If the case goes to court, however, it may be tricky to maintain the anonymity of the customer unless you have legislation such as they do on Guam.

Consider using your web site to communicate information on your revenue protection program. The web site can also be used to allow customers to anonymously report neighbors who are tampering. Several examples of revenue protection information on utilities web sites are included in Appendix E.

The major points of your initial revenue protection communications with your customers include discussions of amnesty programs, the fact that energy theft is against the law, that honest customers pay for what thieves steal, and that thieves can create dangerous situations. If you have an amnesty program, explain how long the program will be offered and how they can apply. Discuss your state law and the penalties a convicted energy thief faces. Dollars lost to energy theft are built into the rate base and honest customers pay for these losses. Not only have energy thieves killed themselves while stealing, theft has resulted in fires and the injury and deaths of innocent people.

## Meter seal and meter control program

For the program to be effective, you must have a meter seal plan in place. Seals and meters must be controlled, and their integrity must be inspected on a scheduled basis. If you are starting a new program, consider replacing all the seals on your system over the next several months. Use different seals that are color coded by account:

- Green seal for an active account
- Yellow seal for an account with a medical emergency need
- Blue seal for a suspect account
- Red seal for an inactive account

Seals and meters must be in a secure area both in the warehouse and the field. The utility should be able to account for the location of each of its meters, at least those that haven't been stolen. Software currently available is very helpful for achieving this objective. Seals should be issued to employees by serial numbers and employees must be held accountable for controlling those seals. All old seals, including demand seals, and out of service meters should be brought to the meter shop for disposal.

If you have contract meter readers, they are probably being paid a flat fee for each meter read. If you ask them to check the integrity on meters, they are going to read fewer meters and make less money in a given time period. If contract meter readers are asked to inspect seals, be prepared to compensate them for the additional time required to complete this activity. Make sure they are not looking for theft alone but other metering problems as well. And be sure that you train them so that they know what to look for and how to report unusual or suspicious situations.

## Testing and cycling of meters

Many utilities are replacing old meters with new ones, often electronic meters or automatic meter reading systems. If your utility is still using mechanical meters, however, remember that they slow down after years in the field and are not recording all the consumption. They should be rotated back to the meter shop every eight to ten years for testing and calibration.

When receiving a shipment of new meters, test at least a sample of each shipment to ensure they are registering correctly. Even new meters are susceptible to problems. Many utilities check every new meter, but this is not always possible.

## Antitampering devices

There are a number of antitampering devices on the market. The best antitampering devices are well-made seals and locks that make tampering more difficult. These will usually help to keep the honest thief honest, but they will not deter the dishonest thief. If you make it impossible to tamper with the meter, the determined thief will use a bypass to save on their utility bill. Decide which seals and locks are best for your system, but also but also keep the awareness level high among employees responsible for detecting energy theft.

## Strike force

A *strike force* consists of utility employees supported by local police. Inspect all the meters in an area where a lot of energy theft is suspected. A strike force team can consist of two utility employees—a meter technician and a revenue protection investigator—and a uniformed police officer. The strike force may operate during the week when many customers are not home, or they may operate on the weekend with the intent of confronting customers caught stealing. When operating during the week, the objective is to find energy theft, process the crime scene, remove the meter, and leave the customer a notice. When operating on the weekend, the objective is to show the customer what you have found, call the office for an estimated back bill, and collect the back bill before restoring service. If service is restored, a new meter will need to be installed because the old meter is now evidence.

Strike forces are not used as much as they were a decade ago for two reasons. First, there is a concern regarding the cost of the overtime for the utility field employees, the employee in the office computing back bills, and for the police officer. The other concern is safety. In some areas crime rates and the potential for violence has increased, and these high crime areas are where you are most likely to find an abnormal number of energy thefts. Strike force activities are used in countries with large amounts of unaccountable losses and a need to make a major impact on the energy theft problem. For example, they have been used in recent years in Jamaica and in Pakistan.

### Reading meters at inactive accounts

We have discussed the need to read meters of inactive accounts monthly. The utility needs to ensure that the account is still inactive and that the meter is still there. The meter should be removed from an account that remains inactive for some time.

There should also be a program in place to read meters of accounts that have been disconnected due to nonpayment and are not reconnected within five business days. Accounts that are disconnected for nonpayment who self-reconnect make up the greatest percentage of revenue protection investigations in most utilities. If energy theft is suspected, inspect the account after dark to determine if lights are on. If so, you may consider visiting the account with a police officer. Show the officer the work order disconnecting the service, as well as evidence that the service has been reconnected illegally, and that who ever is using the service is receiving stolen property. Now it's time to knock on the door. Let the officer do the talking. The customer will not always be arrested. Often the situation is resolved by disconnecting the service again and requiring the customer to come to the utility the next day to pay the bill and arrange for the service to be reconnected, legally.

## Reactive Activities

So much for proactive activities. A suspect situation has been found and now is the time for reactive activities. The fourth component of the program is about to spring into action.

### Possible theft detected

Either a customer has telephoned a tip that someone is stealing or an employee found an irregular meter situation. It's now time to mobilize the investigative team. This may be a team of one if the investigator is a meter technician. If not, the team will include the investigator and a technician. A police officer may also be called, depending on the utility's policies and working relationship with local law enforcement.

## Crime scene secured

The location of an energy diversion is a crime scene and should be treated as such. The investigator must follow the rules of crime scene investigation (discussed in chapter 3) by processing the scene and collecting the evidence according to the rules of law. If the customer is present, it is preferable to have a police officer at the crime scene. The officer will handle the customer while the utility employee completes the investigation.

## Initial photographs

If there is any reason to believe that tampering or diversion have occurred, take photographs of the scene as you found it when you arrived, even if there doesn't appear to have been energy theft. As you uncover the scene, you may find that theft has occurred and has been cleverly concealed. More photographs will be taken as you remove the meter and find additional evidence. Evidence such as jumpers in the meter should be photographed before they are removed. When you complete the crime scene investigation, photograph the scene as you leave it, either with a new meter in place or with a cover on the meter base, depending on your utility's policy.

## Examine the scene

Crime scene investigations must be systematic and detailed. Start from the outside and work in toward the meter. Start looking for evidence from the time you exit your vehicle, not when you're standing in front of the meter. Systematically inspect the secondary line from the transformer to the meter, and look for additional wires connected to the transformer. One clever thief ran a secondary service from his overhead transformer, carving a notch down the side of pole, running the wires through the notch and then filling in the notch with plastic wood. He then ran an unmetered underground service from the base of the pole to his house. The free service was undetected for almost 20 years. If the service is an underground, look for indications of digging along the service line.

Be systematic in your investigation. Inspect the scene as you find it. Look around the meter seal and locking ring for indications of tampering and then remove the meter. Look behind the meter socket for unusual wires or other problems.

## Collection of evidence

Evidence must be collected according to the rules of evidence discussed in the next chapter. Each piece of evidence must be identified at the scene, entered into your field notes, and then secured. It must be secured at the crime scene, in transit to the utility, and while held at the utility.

## Complete your field notes

Before you leave the scene, complete your field notes. Some investigators do this in writing, whereas others use mini tape recorders. Note all the details of your investigation including your gut feelings. Have you seen this method used elsewhere in the neighborhood? Does this customer have relatives on your system? You will follow up on your gut feelings later in the investigation. Before you leave the scene, take a photo of exactly how you left it. Then back up approximately 10 yards and take another photo of the meter and the side of the building. Then back up another 10 yards and take a final photo. This will help to confirm the location of the investigation if it is in question later. When it comes to field notes the old adage, "a short pencil is better than a long memory" is a truism. Either write down or record everything you see at the crime scene.

## Disconnect service?

Should you disconnect the service at a location where energy theft occurs? That depends. If the service is inactive or has been disconnected, then it should be disconnected. But what if this is an active account that has been stealing part of the energy consumed but paying the bill for that which has been metered? In some cases courts have ruled that disconnecting the active customer is to deny this person due process. In other jurisdictions this is not a problem. You can always disconnect an active service because of a concern for the safety of the customer. If someone has tampered with the meter, you don't know what else has taken place. In these cases the customer is required to have an electrical inspector or an electrician certify in writing that the service is safe before service is restored.

## Notify the customer

If the service has been disconnected, the customer needs to be notified. If not, then you may leave a door hanger notifying the customer that a problem was found at the meter. Indicate that they need to contact the utility as soon as possible.

## Back at the office

Complete your investigative report when you return to the office. Make sure that the report is secure either in a locked filing cabinet or on a secure computer file. Enter the information found in your evidence log and secure it in a locked container or room. Access to evidence must be controlled at all times. Determine the back bill and other charges related to the investigation. Test the meter to determine how much the tampering affected registration or if diversion was used, figure out how much the meter would have registered to correct amount it had been metered. Make sure that revenue protection supervisors and managers are informed of the investigation and its progress.

## Contact with the customer

All contact with the customer, telephone calls and meetings, should be noted in your investigative file. Note the date, time, persons present, location, and a summary of what was discussed. Also include details of any agreements that were reached during these communications.

The primary objective of a revenue protection program is to prevent revenues from being lost. Its second objective is to recover revenues that have been lost. Recovering these revenues requires a systematic approach to the reactive activities discussed.

## Developing Policies

Your first proactive activity was to develop policy and procedures for your utility's revenue protection program. Now is the time for the fifth component of your program. One of the first decisions is to determine if you will have full-time revenue protection employees. As discussed in chapter 1, an International Utility Revenue Protection Association (IURPA) study found that among the association's members responding to the study, there was one investigator for approximately every 52,000 meters. However, some utilities with 20,000 to 25,000 customers find that they can cost justify a full-time investigator. In these utilities the investigator duties include the following:

- Conducting diversion and tampering investigations

- Conducting customer fraud investigations

- Investigating disconnects for nonpayment who are not reconnected within five business days

- Investigating other metering problems that result in a loss of revenue

Your policy should assign responsibility for revenue protection to a specific department, but the other departments involved should have their responsibilities outlined. Procedures to be followed when a suspect situation or actual theft is discovered should be covered in the procedures. Sample policy and procedures are provided in Appendix B.

## Summary

Use the information in this chapter to organize your program and assign responsibilities within the utility. Decide who will be on your initial task force and when you will conduct your monthly meetings. Once the program is established, these meetings will be scheduled quarterly.

If you are going to assign or hire full-time investigators make sure they are trained. If you are using current employees who have revenue protection as an additional duty, they also need to be trained. And use the same employees for all your investigations.

# Recommended Actions

Make sure that each of the five components of the revenue protection model is being addressed by your program:

- Who is responsible for each of the elements in each component?
- Which department is responsible for revenue protection?
- What are the roles and responsibilities for the other departments within the utility?
- Will you have full-time investigators or assign revenue protection responsibilities to current employees as an additional duty.
- Have these employees been trained to handle revenue protection investigations?
- Which proactive activities will you decide to use?
- Are you prepared to complete all the reactive activities? These are not optional.
- Do you have a revenue protection policy and procedures in writing?

# 3
# Crime Scene Investigation: Theft of Service

## Detecting Theft of Service

The front-line defense in combating theft of service is meter readers and other field personnel. They are the utility's eyes and ears. Looking for and reporting actual or suspected meter tampering and current diversion should be in their position descriptions. Some utilities also have reward systems to encourage field personnel to detect utility theft. Here are some examples of reward programs:

- The utility pays $25 for every report that results in a potential revenue recovery. This includes metering problems other than meter tampering or current diversion such mechanical problems or damage resulting from lightning.

- The field employee receives a percentage of the potential revenue recovery. For example, the employee receives 5% of the back bill with a minimum of $25 and a maximum of $500.

- The employee receives points for the detection of revenue protection cases. Points are also awarded for accuracy in reading, safety, and so forth. At the end of each month the employee with the most points receives a cash bonus, such as $100. At the end of the year the employee with the most points receives a more substantial annual cash bonus, $1,000 for instance.

- No money is awarded, but the utility presents plaques annually for recognition in a number of revenue protection areas. These include, for starters, the most number of cases detected, the largest single revenue recovery, and the most potential revenue recovered. The awards are presented at a breakfast or luncheon.

Many managers do not believe in a detection rewards system because finding utility theft is part of the field employee's job. Although that is true, somehow employees seem to do their jobs more effectively if there is a little extra incentive. Also, incentive based on a percentage of the lost revenue should be paid for the total back bill even if the money is not collected. The field employee has control over the detection but not the actual revenue recovery. Checks can be presented monthly at a tailgate meeting. Keep the incentive checks separate from the paychecks so that all field employees are reminded monthly that detection is important.

The computer department will generate the utility's exceptions report each month. Several formulas for computing the report have already been discussed. The report includes consumption that is higher or lower than expected and all meters where a multiplier or constant is used in the billing. The report should be reviewed monthly by metering, meter reading, billing, and revenue protection. Exceptions that cannot be explained should be inspected by service personnel during the next month. Glaring exceptions should be inspected as soon as possible.

Your other eyes and ears in the field are customers. Provide a telephone number and an e-mail address or message function on your web site, where customers can anonymously report meter tampering or power diversion. Make sure customers understand that the report is confidential and that it will be investigated.

The revenue protection investigator also plays a key role in the detection of theft of services. If the investigator identifies three people who are stealing services and all work at the same place, find out who else works with them and check their meters. If you identify several people in the same family stealing, identify other family members and inspect their meters. When you find a thief, often someone told him how to do it and he told someone else how to do it. You may have found the tip of the iceberg. A good investigator knows how to get below that tip.

Remember that your objective is greater than just the detection of tampering and power diversion. It should also detect

- Unlisted meters
- Defective meters
- Cut or missing seals
- Inaccessible meters
- Change-out meters
- Accounts with no consumption
- Demand meters with zero consumption
- Fraudulent reconnects

The information in figure 3–1 is from an actual monthly revenue protection report. Based on what we just discussed, do you see any problems with the report?

This report is for meter tampering and current diversion cases only. It does not include other metering and billing problems identified by the revenue protection program or fraud cases that were investigated.

Who are the utilities' eyes and ears in the field? Meter readers! Yet on this monthly report meter readers reported one residential potential theft and one commercial potential theft. Neither was confirmed as meter tampering or power diversion.

There were 43 confirmed residential thefts during this month. The utility collected $3,416.88 from these accounts; $4,601.08 was "unbillable" because the customers could not be located. This is an average of $186.65 per residential account. There were six confirmed commercial account investigations resulting in a potential recovery of $20,638.91. This is an average of $3,439.82 per account. Where is the real potential for revenue recovery?

## RESIDENTIAL ACCOUNTS

| Origin | Thefts Reported | Thefts Confirmed | Billed | Unbillable** |
|---|---|---|---|---|
| Security | 0 | 0 | $0.00 | $0.00 |
| Customer Service | 17 | 7 | $303.00 | $1,377.33 |
| Meter Maintenance | 15 | 11 | $789.55 | $984.00 |
| Reconnect | 13 | 5 | $45.00 | $1,576.10 |
| Credit and Collections | 54 | 18 | $1,521.71 | $663.65 |
| Meter Readers | 1 | 0 | $0.00 | $0.00 |
| Other | 3 | 1 | $757.62 | $0.00 |
| Monthly Total | 103 | 43 | $3,416.88 | $4,601.08 |

## COMMERCIAL ACCOUNTS

| Origin | Thefts Reported | Thefts Confirmed | Billed | Unbillable** |
|---|---|---|---|---|
| Security | 0 | 0 | $0.00 | $0.00 |
| Customer Service | 0 | 0 | $0.00 | $0.00 |
| Meter Maintenance | 0 | 0 | $0.00 | $0.00 |
| Reconnect | 1 | 0 | $0.00 | $0.00 |
| Credit and Collections | 28 | 6 | $20,638.91 | $0.00 |
| Meter Readers | 1 | 0 | $0.00 | $0.00 |
| Other | 0 | 0 | $0.00 | $0.00 |
| Monthly Total | 30 | 6 | $20,638.91 | $0.00 |

Fig. 3–1. Monthly revenue protection report

The credit and collections department deserve recognition for the success of this program, along with the efforts of the single investigator working in a service area with over 300,000 meters. The meter readers, who are employees not contract workers, need to get with the program.

# Principles of Investigation

The first principle of investigation, other than having an organized revenue protection program, is the selection of the investigator. A good investigator is detail oriented, methodical, and has a sense of humor. The investigator must be good at working with people, especially people who are caught stealing.

Because utility theft is a crime and the place where it occurs is a crime scene, the investigator must know how to conduct a crime scene search, take photographs, collect evidence, and document the processing of the crime scene. For this reason, some utilities hire former law enforcement officers to conduct revenue protection investigations. These people are trained, experienced investigators and they have experience in working with people who have committed crimes.

Other utilities prefer to train metering personnel to conduct revenue protection investigations because they have the technical expertise required. They must know how to process a crime scene, as well as how to collect and maintain the custody of the evidence. The lack of properly trained employees can lead to problems.

In the late 1970s a utility in the midwestern United States made a commitment to institute a meter tampering program without first determining what the requirements of the program should be or what they should do when they caught someone stealing. In one of their first cases where a customer had tampered with his meter, they collected the meter and evidence and left the customer in the dark. The meter and other evidence remained in the open bed of the utility's truck for several days. When it was apparent that the customer was not going to pay the back bill (he and his wife had moved in with a relative), the evidence was placed on a table in the manager's office. First mistake, the custody of the evidence had never been established.

A second mistake occurred when the customer would not respond to the utility's demands. The manager of the utility swore a warrant for the customer's arrest on charges of theft of services. Although the warrant was sworn in the morning, it was not served until later that night. Because the county did not have a night magistrate, the customer spent the night in jail and then could not make bail until the next morning.

When the case went to court, the defense attorney representing the customer challenged the evidence because its custody had never been established. The judge agreed. The case was thrown out of court, and the customer went home to a house that now had lights.

The customer went on to sue the utility for malicious prosecution and false arrest. The court awarded him $365,000. The utility appealed the award saying the "judge erred in his instructions to the jury" before they made their finding. The appellate court agreed and the award was overturned.

The customer sued again and was awarded $585,000. This time there was no basis for an appeal. Then the customer's wife sued the utility. The night he spent in jail severely traumatized the customer and he has not been able to perform his "husbandly functions" as well as he used to. She settled out of court for approximately $250,000. The customer and his wife received $835,000 from the utility after they caught him stealing. The custody of the evidence *must* be established and maintained.

The second principle of investigation is the processing of the crime scene. This includes the initial arrival at the scene, crime scene photography, and the collection and processing of the evidence. Each of these activities will be discussed in detail in this chapter.

The next principle is to know how to follow up on the crime scene investigation. The name the utility has for the account might not be the real name, and the investigator needs to find out who the actual owner or renter is. The investigator may find several locations where the same innovative approach was used and realize there is a fixer in the area. We need to find out if these customers have had electrical work completed recently and, if so, by whom?

Knowledge of human factors is an important principle. The investigator should understand the difference between honest and dishonest thieves. It is also important to know that occasionally when conducting an investigation, someone will lie to you. A good investigator knows how to detect a lie and how to respond to it. The investigator must also understand that some people are mean and dangerous. These people should not be approached unless accompanied by a law enforcement officer.

Another principle is to understand case assignment priorities. An investigator who gets below the tip of the iceberg is going to find more cases than one person can work. The cases that will be most productive, in terms of revenue recovery, usually receive priority. High-profile cases, however, such as when you catch a prominent person in the community, may receive an even higher priority.

Knowledge of the rules of evidence is a required principle. Failing to understand this principle cost the utility we discussed $835,000. The thieves used that money to retire and build a new house.

The final principle of investigation is case supervision. Someone needs to monitor the investigator's activities to ensure that this person does not become too focused on certain aspects of a major case while ignoring others. The supervisor also makes sure that the case priorities established by the investigator are appropriate for the goals of the utility's revenue protection program.

# Rules of Evidence

In general, evidence used to prove the elements of a crime can be split into two types: testimonial evidence and physical evidence. The testimonial evidence would be a witnessed account of the crime scene. The physical evidence would refer to material items collected at the crime scene. Evidence is presented to prove or disprove the facts of the case. Evidence is used to

- Prove that a crime has been committed.
- Establish the key elements of the crime.
- Link a suspect with the scene or location.
- Establish the identity of the suspect or victim.
- Corroborate verbal witness testimony.
- Exonerate the innocent.

There are six categories of evidence. To examine them, let's consider the following scenario. A service person arrives at a residential location to inspect a meter. When the employee pulls the meter, he finds wires running from the potential circuit through a hole in the meter base into the house and notifies the revenue protection investigator. The service person remains at the scene until the investigator arrives and confirms what has been reported. The investigator rings the doorbell and a woman answers. He explains there may be a safety problem at the meter and he would like to come into the house to check the back of the meter base. The woman says that she is alone and doesn't feel it would be proper for him to enter the house. He agrees but explains that they must turn off the electricity. He gives her a business card and suggests that her husband telephone him. The investigator goes back to the meter bases and pulls on the wires running off the potential. There is resistance but they finally come free. The other ends of the wires are stripped. They collect the meter, the wires and the seal as evidence and place a meter cover and lock on the service. Photographs are taken throughout the field investigation and the custody of evidence has been established.

**Direct evidence**

This is eyewitness testimony. The witness can only testify to what his senses took in. In our example, the meter reader would present direct testimony if the case went to court. With regard to the stripped wires, he could only testify as to what he actually saw, not to conclusions as to why the wires were stripped at the other end.

**Real evidence**

This is the physical evidence collected at the crime scene. In our example it includes the seal, meter, and wires.

**Circumstantial evidence**

Circumstantial evidence is that which would lead a person of normal intelligence to suspect that a crime has been committed. A cut seal, for example, is not hard evidence of tampering, but it may be an indication of tampering.

## Opinion evidence (expert testimony)

Unlike someone presenting direct evidence or testimony, when an expert witness takes the stand, the attorney who called the witness first establishes the credentials and asks the court that the witness be considered an expert in this particular field. Credentials are established by reviewing the person's résumé and experience as they relate to the case. The attorney for the other side has the right to question the witness further before the judge makes a decision. Once qualified as an expert witness, the person can present opinions. In our scenario, the investigator would be qualified as an expert witness. When asked why the wires were stripped at the other end, the investigator could render an opinion that there was probably an on and off switch on the inside of the house allowing the customer to control when the meter was registering the electricity used.

## Documentary evidence

Documentary evidence includes the customer's application for service and billing records. It also includes the investigative report and photos of the crime scene.

## Scientific evidence

This includes fingerprints, DNA samples, and the like. Scientific evidence is usually not used in utility theft cases. The closest we come to this is the testing of the meter, but the meter test report is usually presented as documentary evidence. The rules of evidence require that physical evidence must be identified and properly collected at the crime scene. The evidence should be fully described in the investigator's field notes, including any identification numbers. Describe how the evidence was found, its relationship to the case and note how it was photographed. Photograph the evidence as found. In our example, the investigator would have photographed the wires going through the back of the meter base before and after they had been removed. A smart investigator would have also placed the wires on the ground next to a ruler and taken a photograph that shows their exact length.

Once the evidence is identified, photographed, and collected, it should be tagged and bagged at the scene. The investigator should use the same evidence tags and bags that are used by local law enforcement. Professional tags and bags require you to record the information necessary to establish the custody of the evidence. Ask your local law enforcement agency where they purchase them. The use of investigative tags and bags helps to maintain the custody of the evidence by providing a record of

- Who had contact with the evidence
- The date and time the evidence was handled
- The circumstances for the evidence being handled
- What changes, if any, were made to the evidence

Evidence entered on the tag or bag includes

- Description of the item
- Case file or identifier
- Date
- Location of collection
- Collector's name and identifier
- Brand name
- Any serial number or other identifying information

A number of evidence tag and bag suppliers can be found by searching the Internet using the term *evidence bags*. The search will help you locate suppliers such as Galls Law Enforcement Supply, Arrowhead Forensic Products, and Security and Safety Supply Company.

Once the evidence is tagged and bagged, it must be locked in the vehicle in the field. Don't leave evidence bouncing around the back of a pickup truck. It must be locked in the vehicle or in a locked container on the vehicle. Access to evidence must be controlled at all times.

Maintain an evidence log at the utility. Again, ask your local law enforcement agency for the name of a supplier or for blank copies of the log pages they use. Every piece of evidence must be logged into the log book. When evidence leaves the evidence locker, note the time and date out, who had possession, and the time and date it is returned to the locker. Access to the evidence locker is limited and controlled. Investigators have keys and there should be a key in the utility vault.

Let's assume that a meter was found to be manipulated. It was collected, placed in an evidence bag, and the bag has been secured in the evidence locker to be tested. The investigator will log the meter out of the locker and take it to the meter shop and remove it from the evidence bag. The investigator will stay with the meter while it is tested. When the test is complete, the investigator will place the meter in a new evidence bag along with the original bag. He will then complete the information on the outside of the new bag to establish the continued custody of the evidence. The meter will then be logged back into the evidence locker. The custody of the evidence must be established and maintained. No excuses.

## Processing the Crime Scene

A crime scene investigation has three basic stages: scene recognition, scene documentation, and evidence collection. An organized crime scene search ensures that the sequence of established protocols is followed:

- Thorough and legal search conducted
- Scene processed expeditiously and without compromise
- Scene properly documented
- Proper methods and techniques for evidence recovery applied
- Resources and equipment accurately and knowledgeably used
- All pertinent evidence recovered
- Evidence properly handled and packaged
- Proper safety precautions followed

The crime scene search must be systematic. It is usually conducted from the outside, which means the search begins as you drive up to the scene. Be alert for evidence of utility theft by checking the secondary lines or where the underground connection goes from the pad transformer to the meter. Determine if there are vehicles in the driveway or garage indicating that the customer is present. If the location is a business, what activity is taking place and who is present? You should have the initial information or report that brought you to the scene and, if possible, the name of the customer on record, a history of consumption, and other information on the account. Review these before you leave the vehicle.

The investigator must also have the equipment needed to process the crime scene. In addition to tools used to remove and inspect the meter and service and dog repellent, investigative tools include

- Camera
- Voice-activated mini tape recorder
- Writing pad
- Binoculars
- Flashlight
- Tape measure
- Ruler, 12 or 18 inches
- Evidence tags and bags

As a utility employee, you have limited access to the property. Utility employees can inspect anything along the right-of-way, at the utility's equipment, and where the meter reader normally walks. The investigator does not have the right to be elsewhere on the property or looking into the windows or other buildings. A search warrant is needed to inspect areas other than those listed. The search warrant will be obtained either while working with the police or directly from a magistrate or judge. If a warrant is obtained, it will state specifically the areas it authorizes you to search. Although laws differ in different jurisdictions, it is recommended that you never serve a search warrant without a law enforcement officer present.

Let's discuss crime scene processing by beginning with the big picture, the basic stages of the search using a list adapted from the California Commission on Peace Officer Standards and Training, and then examine a specific checklist developed by a major electric utility in the United States. There are 10 basic stages in a search:

1. Approach, secure, and protect the scene. Assuming the utility employee approached the scene alone, it is secured when you arrive. If the customer is present, you may want to have a law enforcement officer with you to protect the scene as you conduct your investigation. Are there unusual footprints or other possible evidence on the ground near the meter?

2. Initiate a preliminary survey and determine scene boundaries. Determine if the investigation will be limited to the area around the meter or will include other areas of interest. If other areas are considered, do you have a right to be there or do you need a search warrant?

3. Evaluate physical evidence possibilities. In part, this will be determined by the method used. If tampering has occurred, the meter and seal will be evidence. If a bypass was used, it is the primary evidence, however, collect the seal and meter that was present when you arrived at the scene as well.

4. Prepare a narrative description. If you use a mini tape recorder you can record an ongoing narrative description. Otherwise, outline the steps you take in writing and list what you find so that you can complete the narrative report at the office.

5. Depict the scene photographically. Take photos when you arrive at the scene, as you investigate and disassemble the scene, and as you leave the scene. If there is a vehicle in the drive, take photos of the vehicle and license plate. Remember to take several photos of the building as you leave.

6. Prepare diagram and sketch the scene. We normally do not need sketches at theft of service investigations. A possible exception occurs when an employee discovers an unusual situation, for example an inverted meter, and when the investigator arrives, the situation has been corrected. Ask the person who discovered the scene to sketch what was found. Make sure the person dates the sketch; notes the location, date, and time the situation was discovered; and signs it.

7. Conduct a detailed search. Refer to the specific list that follows this general discussion. If you are new to theft-of-service investigations, you may want to develop your own investigative checklist.

8. Record and collect physical evidence. Photograph the evidence in place. Tag it and bag it at the scene. Record all the details regarding the evidence in your field notes or narrative.

9. Conduct a final survey. Take one last look around before leaving the scene. Have you identified and collected all the evidence? Do you have all your tools and investigative equipment?

10. Release the crime scene. The scene is released when you leave. Be sure to take a photograph of exactly how you leave the scene.

Now let's look at the 15-item checklist developed by an electric utility in the United States. The checklist was developed to train new investigators to make sure they conducted a detailed search. It could be modified for use at your utility.

1. Identify yourself to anyone present and say that you are conducting a "safety inspection."

2. Visually inspect the transformer and secondary line or underground route. Use the outside-in method when approaching the meter and look for evidence around or under the meter.

3. Photograph the scene exactly as it is found.

4. Verify the meter serial number and record the meter reading.

5. Check the seal or lock for tampering.

6. Check the meter ring to ensure that is properly installed. Look for pry marks or other evidence of tampering. Be sure to check the rivets.

7. Check connection at entrance point.

8. Photograph any irregularities found.

9. Apply load to meter and check for disk rotation.

10. Pull main breaker or switch to the off position and remove meter to check

    a. Cover for hole

    b. Internal seal

    c. Potential clip

    d. Unusual wires

11. Check meter socket for

    a. Current diversion wires (jumpers)

    b. Test line side of socket for current

    c. Test load side of socket for current

    d. Check jaws for arching or scratches

    e. Look for unusual wiring in socket

    f. Look for anything else that doesn't belong there

12. Set the meter and record meter serial number and reading.

13. Collect, mark, and log all evidence.

14. Complete field notes.

15. Photograph the scene as you leave it.

# Documenting the Crime Scene

Although each utility has its own method used to document the crime, it is important that all notes and reports should be presented in the chronological order of the events and should not include opinions, analysis, or conclusions. These will be included in the investigation report, which includes the crime scene report. The investigation report will also include the meter test report and the estimated or actual back bill.

The narrative crime scene report includes five sections:

1. Summary
2. Scene
3. Processing
4. Evidence collected
5. Pending

The summary includes the details that led to the investigation. For example, "Meter reader Henry Smith reported a broken seal and what appeared to be low consumption at this address. Unexpected low consumption was confirmed by an analysis of the billing over the past twelve months." The summary provides the probable cause for initiating an investigation. It is brief and may refer to other sections in the report.

The processing section explains what you did at the scene during the investigation. List the location, time, date, persons present, and other basic details such as contact with the customer and, if it is relevant, weather or lighting factors. List your actions at the scene step by step in chronological order. The processing section is supported by relevant crime scene photographs.

In the evidence section list the evidence collected, how the custody of evidence was established, and how it is being maintained. Use photographs of the evidence to support this section. State the relevance of each piece of evidence to the investigation and to the method used to steal services.

The pending section outlines the tasks that need to be completed following the crimes scene investigation. These include confirming the owner at the address, computing the back bill, testing the meter, and other investigative activities depending on the nature of the case and the outcome of the crime scene investigation.

Photographs taken during the crime scene are usually taken with a 35 millimeter, Polaroid, or digital camera. Most crime scene photography used 35 millimeter cameras in the past but that trend is changing. Using a Polaroid guarantees you know exactly what photos you have when you leave the scene.

Many utilities and law enforcement agencies now use digital photography. Until recently the admissibility of digital photography was questioned because it is easy to modify or morph a digital image. Most U.S. states now have laws on the admissibility of digital photography in court and you should check with your attorney or the prosecutor's office regarding the requirements in your jurisdiction. By taking the following steps, you help to ensure that your digital photographs are admissible:

- Develop a Standard Operating Procedure (SOP) or utility policy on the use of digital imaging. The SOP should state when digital imaging will be used; how to establish the chain of custody, image security, image enhancement; and the storage of digital images.

- Preserve the original digital image. Record the original image to a hard drive or to a CD or DVD. Digital images should be preserved in their original file formats. When you save a file in some formats, the image may be subject to *lossy compression*. This refers to data compression techniques in which some amount of data is lost. Lossy compression technologies attempt to eliminate redundant or unnecessary information. Most video compression technologies, such as MPEG, use a lossy technique. If lossy compression is used, critical image information may be lost and artifacts produced as a result of the compression process.

- If images are stored on a computer workstation or server, and other individuals have access to the image files, make the files read-only for all but the investigator.

- If an image is to be analyzed or enhanced, the new image file should be saved as a new file with a new file name. The original file must not be replaced or overwritten with a new file.

Some utilities are using video recording of crime scenes. Newer cameras have the option of both digital still photography and videotaping capabilities. Videotaping is valuable for showing an overview of the crime scene, but there are some basic rules to follow:

- Start the videotape with a brief verbal introduction. The introduction includes the date, time, location, persons present, why this service is being investigated, and any other introductory information.

- Videotape the crime scene after the introduction with the audio turned off. Unfortunately, things are said at some crime scenes that are best not preserved for future listening. If you record the audio and remove it later, however, you have tampered with the evidence and the video will not be admissible if you go to court. Turn off the audio when videotaping a crime scene.

- Begin videotaping the crime scene using the outside-in approach. Tape a general overview of the scene and the surrounding area.

- Videotape evidence in place before it is collected. Videotape the scene exactly as you leave it.

- Use slow camera movements, especially when panning and zooming.

## Summary

Investigators must be trained. They have to have some technical expertise and a lot of investigative expertise. They must know how to process the crime scene and how to establish and maintain the custody of the evidence. Investigators must know how to deal with customers who are stealing from the utility without being confrontational and how to work with the customer to recover lost revenues.

## Recommended Actions

- Be sure that the detection of revenue protection is included in the position descriptions of field personnel.

- Decide if your utility will implement a detection reward system for field personnel. If so, determine how the system will work and how the rewards will be disseminated.

- Develop a monthly activity report for your revenue protection program. An example is found in this chapter. The report should be posted on the bulletin boards in the utility to remind employees of the importance of the revenue protection program.

- Develop procedures to establish and maintain the custody of evidence. These should include the use of investigative tags and bags and an established evidence log at the utility.

- Review the list of evidence tools presented in this chapter. Be sure that your investigators have the tools they need. Make sure they know how to use the cameras purchased for investigations and to preserve photographs, particularly digital images, as evidence.

- Be sure that your investigators are trained to process a crime scene and to collect and maintain the custody of evidence.

- Consider developing an investigator checklist. This not only helps to train new investigators but becomes part of the crime scene report and shows what the investigator inspected during the field investigation.

# References

Byrd, M. 2000. Duty description for the crime scene investigator. http://crime-scene-investigator.net/dutydescription.html (accessed July 5, 2004).

Byrd, M. Proper tagging and labeling of evidence for later identification. http://crime-scene-investigator.net/tagging.html (accessed July 5, 2004).

Byrd, M. 2000. Duty description for the crime scene investigator. http://crime-scene-investigator.net/dutydescription.html (accessed July 5, 2004).

California Commission on Peace Office Standards and Training's Workbook. 1993. Organization and procedures for search operations. http://crime-scene-investigator.net/respon3.html (accessed July 5, 2004).

Staggs, S. B. 2001. The admissibility of digital photographs in court. http://crime-scene-investigator.net/admissibilityofdigital.html (accessed July 5, 2004).

Staggs, S. 1997. Techniques for crime scene video taping. In *Crime Scene and Evidence Photographer's Guide*. http://crime-scene-investigator.net/video.html (accessed July 5, 2004).

# 4
# The Fraud Investigation

## The Fraud Problem

At 9:30 this morning the utility disconnected the service of Jerry Smith at 14 Main Street for nonpayment. At 2:30 this afternoon Louise Wilson applied for service at 14 Main Street. No, she does not know Jerry Wilson and she just moved to that location this afternoon. She believes that the former tenant moved out several days ago. Because they want to provide the best customer service possible, the utility had Ms. Wilson's service connected by 5:00 p.m. the same day. The utility never found Jerry Smith and eventually wrote off his balance as a loss. Three months later Louise Wilson skipped owing the utility a final bill. That bill was also written off as a loss.

What is wrong with this scenario? The major problem is that this happens every day at utilities around the world. People play the *name game* where the same customer you disconnected is reconnected using a new name, or they play *revolving door game* where another person at that location claims they just moved in and has the service activated in their name. Regardless of the game played, the outcome is the same. The customer eventually skips, leaving the utility and its honest customers with the loss.

That fact that customers can have a utility service put in the name of one of their children, a dead person, or even the name of a pet, suggests that some utilities need to do a better job of knowing who the customer really is before connecting the service. And if most of the money lost to skips is being written off, then the utility is not doing what it should to find these people and collect the lost revenues.

Many utilities have improved their customer application systems by conducting credit checks of all new customers. At a minimum, the credit check will confirm that the name and social security number (*national identity number* outside of the United States) matches. If a customer skips owing a final bill, these utilities enter the fact that the customer owes a back bill on his credit report. In some cases, they are able to use credit reporting services to locate the customer and attempt to collect the money.

In the age of facsimiles and the Internet, utilities encourage customers to apply for service using these media rather than coming to the office. If the utility allows customers to complete their application for service electronically, without ever visually checking the customer's identification, then even if the utility runs a minimal credit check, they are still not certain that the customer is who he says he is. Fraudulent customers are beating the system by taking advantage of one of the fastest growing crimes, identity theft. They use a real person's identity to obtain utility services. Then when they skip, the utility attempts to collect its lost revenues from the identity theft victim.

Before we deal with the specifics of fraud against the utility, let's discuss fraud in general. What constitutes and what motivates this category of crime. Fraud occurs when all the following elements exist:

- An individual or an organization intentionally makes an untrue representation about an important fact or event.

- The untrue representation is believed by the victim, in our discussion the utility

- The utility relies on and acts on the untrue representation.

- The utility suffers loss of money and/or other property as a result of relying on and acting on the untrue representation.

Fraud can be committed for the benefit of an individual or, as we shall discuss, a business or other organization. When an individual commits fraud, the benefits and gains may be direct or indirect. Indirect gains include promotions, bonus, or influence and are usually the consequence of employee fraud. When an organization commits fraud, the benefits and gains are usually direct in the form of financial gain.

The motivation for fraud is often explained using the *fraud triangle*. As shown in figure 4–1, the three elements of the triangle are opportunity, pressure, and rationalization. To decrease fraud losses, the utility must understand which of these elements it can influence.

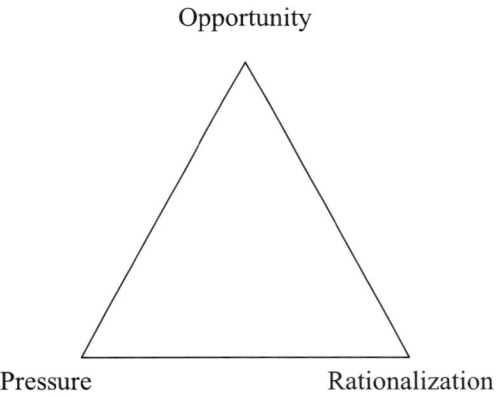

Fig. 4–1. The fraud triangle

The utility has some control over opportunity. It can be more effective in checking the customer's identification before connecting service. It can monitor problem accounts. And when a customer is disconnected for nonpayment, the utility should be more suspicious when a so-called new customer applies for service at that location, often on the same day. The utility should remove as much of the opportunity to commit fraud as possible.

The utility also has some control over the pressure element. Very few industries work as well with customers who are experiencing financial problems as the utility industry. If a customer has an honest reason for not being able to pay the bill, such as illness in the family, many utilities will arrange a payment plan for the past due amount. Not only will the utility make payment arrangements, but it will counsel the customer on how to use less of its product. Not many industries have similar policies.

The utility has little control over the rationalization element. The rationalization may be that the customer needs to reduce the bill so that he can afford a better lifestyle. He may think, "Everyone else is doing it, so why not me?" or "It's only a little money and the utility can afford it." If there has been a recent rate increase, the customer may use that as the rationalization for defrauding the utility.

Although our emphasis is on customers defrauding the utility, it is interesting to review the results of a utility fraud survey of security directors in the industry. These were their major concerns:

- Theft of cash, lapping
- Theft of materials and supplies
- Theft of service
- Expenses reports overstated
- Payroll related
- Collusion between employees and vendors
- Kickbacks
- Erroneous/improper contractor billings
- Personal use of company equipment
- Credit card misuse
- Personal purchases on purchase orders
- Use of company telephone for personal calls
- Theft by check
- Collusion between employees and customers

# Categories of Customer Fraud

Residential and commercial customers are utility fraud threats. Let's discuss these threats separately.

## Residential fraud problems

**Fraudulent application for service.** This includes customers who play the name game and the revolving door game. The utility doesn't have the correct name on the account to begin with, which makes it difficult to collect when the customer leaves owing a final bill. As we will discuss, fraudulent customers are responding to improved utility procedures to combat this problem by engaging in identity theft.

**Theft of service.** Meter tampering and current diversion are categories of fraud, although these are physical acts rather than behavioral. Some customers who are rightfully afraid to tamper with their meter figure that you can't get electrocuted by playing the name game.

**Skips.** Have you ever had a customer leave your utility without paying the final bill? It happens every day. What do you do about it? In some cases, the utility makes no effort to find the person and collect the bill. In other cases, utilities are very aggressive in developing and pursuing a skip tracing program.

**Bad checks.** Bad checks are a problem for any business that deals with consumers, including utilities. In some cases the utility's policies actually encourage bad checks by accepting checks from customers who have bounced them at the utility in the past and by having bad check charges that are too low. An ineffective disconnect policy will also contribute to the problem.

## Commercial fraud problems

**Fraudulent application for service.** A business fraudulent application for service is different from a residential application. The business may say it's incorporated when it is not or it may list owners who don't know they're owners—because they have never heard of the business.

**Theft of service.** Again, theft of service by meter tampering or diversion is a physical form of fraud.

**Skips.** Like residential customers, businesses disappear overnight. Inventory is moved and the building is locked. If the utility requires the information it should have obtained before connecting the service, these losses will be minimized.

**Planned bankruptcies.** These are a nightmare for the utility. They are difficult to investigate, and it is almost impossible to recover lost revenues in these cases.

**Bad checks.** Businesses like residential customers write back checks. In fact, if you received a bad check from the owner at his business, expect an occasional back check from the same person at his residence.

**Out of business, back in business.** Yesterday it was Happy Harry's Hardware on Main Street, today it's Happy Harry's Carpets on Maple Street with the same owner. The difference is that Happy Harry's Hardware stuck the utility with its final bill. Now Happy Harry's Carpets wants a new service. Guess what? If Happy Harry stuck you at the last business, chances are he is going to stick you again at the new business.

There are several additional categories of fraud that could be committed by either residential or commercial customers. These are *false claims* and *mail fraud.*

You receive a telephone call from an irate customer who claims one of your line trucks ran over his new flower garden, the one he just spent $900 on. No he doesn't have the receipts for the plants because he didn't know one of your trucks would run over them. You did have a line crew working in the neighborhood today but they don't remember a flower garden. When you go to the customer's house it is obvious that a truck did run over his flower garden but there is a surprise when you arrive. The camera crew from a local television station is there. Mr. Customer is complaining to them about the careless driving of your line crew and the damage it caused to his garden. They want to interview you.

What usually happens in these cases? The utility pays for Mr. Customer's garden. What really happened is that Mr. Customer came home drunk last night and ran over the garden with his pickup truck.

Actually, it was his wife's garden. This is what is known as a false claim and it happens all the time.

Under federal law, anyone who engages in fraudulent activity and uses telephones, telegraph, and/or the U.S. Postal Service (USPS) to discuss or either send or receive correspondence or documents in furtherance of the fraud, can be charged for felony mail fraud and/or wire fraud. A customer who uses the telephone to give you false information may have violated a federal statute. A person who pays a bill by mail, knowing that the bill is not correct because they are stealing energy could be guilty of mail fraud. Don't expect the federal agencies to get excited about investigating these cases. They are busy going after major criminals. But notifying customers that a fraudulent application for service is a potential violation of federal statutes and that paying a bill through the mail that is fraudulent may also be a violation will help to keep some of the honest thieves honest. The dishonest thieves don't care.

Figure 4–2 shows the three elements the utility should have in place to address fraud problems.

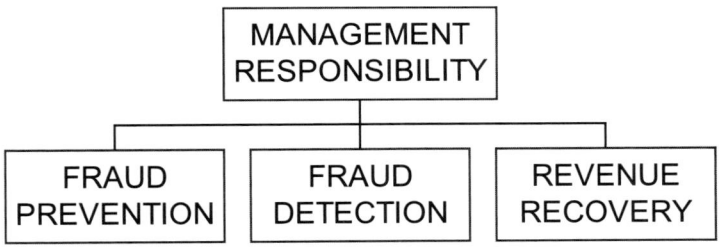

Fig. 4–2. Managing utility customer fraud

When comparing the fraud model with the revenue protection model in chapter 2, it is evident that one of the elements is missing: going to court. This is because fraud cases are often more difficult to prove in court. If a meter has been tampered with, there is physical evidence to prove your allegations. In fraud cases, documents we may or may not have are needed to prove the allegation. In some areas prosecutors turn away almost one-half of the fraud cases referred to their office, including utility customer fraud. However, you may have enough evidence to obtain a judgment in civil court.

# Combating Utility Customer Fraud

### Application for service

There are still utilities around the world that require all new residential customers to complete an application for service at the utility office and to show a photo identification and a credit card with the same name. It is possible to get fake photo identification but it is difficult to use the fake ID to obtain a credit card. The utility will copy the photo identification and attach it to the application. The utility also runs at least a minimum credit check, matching the name and Social Security Number. These utilities have decreased the customer's opportunity to play the name game.

To decrease the customer's opportunity to play the revolving door game, the utility should require a copy of a lease from renters or a proof of ownership from property owners. No exceptions. The utility service must be in the name(s) of the person on the lease. If the landlord signs a second lease with someone else at the property knowing they are attempting to defraud the utility, then the landlord becomes a party to the fraud.

Commercial accounts should be required to provide a copy of their business license when applying for service. They should also provide proof of ownership, for example, sole proprietor, partnership, corporation, or other legal structure. These records are kept with the application for service.

If the utility does allow new customers to apply for service using the Internet or facsimile, it should use at least the minimum credit check approach. However, anyone applying for service at a location that has been disconnected for nonpayment should be required to apply for service at the office, showing photo identification and a credit card.

### Bad checks

There are a number of different forms of check fraud:

- *Forged signatures* usually involve the use of legitimate blank checks, with a false imitation of the payer signature on the signature line. Many cases of forged signatures are perpetrated by someone known to the valid payer. In other cases, the criminal uses blank checks stolen in the mail while being shipped from the printer to the account holder.

- *Forged endorsements* involve the theft of valid checks, which are then enforced and cashed or deposited by someone other than the payee. Marital partners involved in separation or divorce proceedings are a common source for forged endorsements. Forged endorsements can also appear on checks payable to more than one party when one party endorses the check for all parties.

- *Counterfeit checks* are the fastest growing source of fraudulent checks. Check counterfeiters use sophisticated color copiers to copy valid checks. They scan checks digitally and then use software programs to change the valid information. Some of these checks even include the magnetic ink character recognition (MICR) line characters.

- *Altered checks are defined* as valid check stock with certain fields changed. For example, the payee name is changed; payment is made to the wrong person.

- *Check kiting* requires multiple bank accounts and the movement of money between accounts. The check kiter takes advantage of the time required by a bank to clear a check. A check drawn on one bank is deposited in a second bank and the money withdrawn or transferred.

The major check problems for utilities are NSF (not sufficient funds) and account closed. An account closed check requires immediate action. The customer should be telephoned and told the check has been returned. If the phone has been disconnected when you telephone, then you have a problem.

In some cases the utility has an NSF check problem because of its policies. The utility takes a check for payment on the day the disconnect for nonpayment is scheduled and doesn't learn until five to ten days later that it's NSF. Then the utility gives the customer several days to replace the check. What's wrong with this scenario?

First, the utility should have a limit on the number of NSF checks it will accept from a customer before it refuses to accept any future checks for payment. Two NSFs as a minimum, maybe three, and no more checks should be accepted for payment on that account. Second, if the check is returned after the disconnect date, the service should be disconnected immediately. The customer has already received notice of disconnect for nonpayment, that's why he wrote the bad check. Unfortunately in some states the utility is required to give the customer notice before disconnecting the service, even if the customer played the last-minute-NSF-check game to keep the service connected.

What does your utility charge the customer for a bad check? Some utilities have ridiculously low returned check fees; some as low as $5.00. These utilities say they don't want to burden the customer with additional fees. However, customers who write bad checks are not the utility's best customers and the cost of handling those checks is then passed on to the reliable customers who pay their bills on time, with real checks. The utility returned check fee should be the same as is charged by banks in your area. Penalties for bad checks vary greatly. For a list of civil and criminal penalties by state within the United States on the Internet go to http://www.ckfraud.org/penalties.html/.

One of the newer check crimes is *check washing*. Criminals steal real checks, use basic chemicals to remove the ink, and then rewrite the checks in new amounts to new payees. Because these criminals need a lot of real checks, there have been instances where they have robbed the utility's post office box at the end of a weekend and stolen all the checks that were waiting to be picked up on Monday. If this happens to your utility the USPS will conduct the investigation but you will also need to notify your customers.

## Planned bankruptcy

According to the Federal Bureau of Investigation (FBI), bankruptcy fraud is a growing national problem. The number of bankruptcies and associated bankruptcy fraud has grown every year. Approximately 70% of these frauds occur when the debtor files for chapter 7: Liquidation and conceals assets. In the case of business bankruptcy fraud, large amounts of cash or inventory are hidden before filing for bankruptcy. Unfortunately there is not much the utility can do to prevent these losses and even if the perpetrator is convicted, the utility will probably not recover any revenues. Fortunately, although the loss can be substantial, there are very few actual cases. In 2003 the Internal Revenue Service (IRS) initiated 58 criminal investigations across the United States. 92.9% of these resulted in convictions with those found guilty were sentenced to an average of 29 months in prison.

## Out of business—back in business

Back to Happy Harry. Happy Harry closed his hardware store and the utility wrote off his final bill because he said he was a corporation. Now the carpet store is a new corporation. Maybe. If the hardware store was not incorporated, then Harry owes the utility the final bill, which must be paid before the new service will be connected. If Harry was a partner in the previous business, then he owes the utility a percentage of the bill that is the same percentage of his partnership. The utility should know the legal structure of the business because it required proof before the original service was connected.

The legal structure of a business can be confirmed in many states by contacting the Secretary of State's office. This can usually be done on the Internet. Go to your favorite search engine, type in the name of your state followed by "secretary of state." This should take you to your Secretary's web site. Somewhere on the site will be a search capability that will confirm whether the business was incorporated. Some businesses incorporate when they begin but don't pay the annual fee and are not incorporated when they go out of business. Others tell you they incorporated when they never were. Ask for proof of the legal structure of a business with the application for service and always verify if the business was still incorporated if it goes out of business leaving the utility with a bill.

Because Happy Harry has stuck you before, consider a substantial deposit before providing service at the new location. Then never let the bill become larger than the deposit before service is disconnected. Or you could ask him to sign a personal guarantee for the business account. The first time Harry beats the utility out of the final bill, it's his fault. If he does it again, it's the utility's fault.

## Skips

In modern societies, no one just disappears. But the ability to find someone who has skipped owing the utility a bill depends on several factors, the most important of which is the information you record with the application for service. As a minimum you should have the correct spelling of the name and the Social Security Number. If possible, get the date of birth, which would be on the photo identification you copied. The Social Security Trace is the most effective search for locating the missing person. This is actually the top of the credit bureau file containing the name and address. People don't realize that various activities will trigger new address information including filling out an employment application, an application to rent an apartment, or applying for utilities at a new location.

Another way to find skips is through the USPS forwarding address files. In the past, you had to go to the post office, file a form, and pay a dollar to get this information but that has changed. The National Change of Address (NCOA) system is now licensed to database companies, and these companies will sell the information to subscribers online. This is much easier than having to physically go to the post office. Approximately 40 million changes of addresses are filed annually and changes are kept in the data base for four years. The information is updated weekly. Each record contains the postal customer's name along with the new and old address. The NCOA information is often combined in the licensed database with magazine subscription information. So if the savvy skip didn't leave a change of address at the USPS, the name and address may still be in the database. To locate a NCOA licensee, go to the USPS Web site at www.usps.gov. This is a competitive business, so request bids from several licensees.

Most of the utility customer fraud cases that go to court end up in civil court. Often small claims court can be used, depending on the amount of the loss, to secure a judgment against the customer. A judgment doesn't mean that you will immediately collect the lost revenue, but it will show on a person's credit report. Eventually he will call you to arrange to settle the matter.

If you do go to court, it is important that you have complete documentation on the case. You will need applications for service (real and fraudulent alike), a history of consumption, rental or ownership records, notes on contacts with the customer, and other documents relating to the fraud.

# Getting Below the Tip of the Iceberg

Do not be in a hurry to confront the customer that you found playing either the name or revolving door game. If he committed fraud against the utility at this location then chances are he has done it to you before. A complete credit check on the customer should show the previous addresses. Check this address against accounts where you have lost revenue. In many cases the customer you just caught can be identified with two to three other cases and the potential revenue recovery goes from around $300 to almost $2,000.

Be creative in your approaches to combating fraud. A municipal electric utility's customer service offices were moved into a central location with all other utility and city services in the same building: one-stop shopping for anyone needing city services. An employee from the utility was talking to an employee from the public subsidized housing department about the fraud problem, learned that if the utilities are not in the name of the person receiving the subsidy check then that person loses the public housing subsidy forever. The utility began checking names and addresses on accounts where there had been problems and found a number of locations where the service had been placed in a new name after being disconnected for nonpayment, but the person who had been disconnected was receiving the subsidy check. These people received a letter from public housing stating that if the utilities were not in their name by the end of the month they would no longer be receiving subsidy checks, and they would never be eligible again. When they went to the utility they found that it had identified all the locations where this person left owing a bill or played one of the games. The individual was presented with a bill for the combined losses, averaging $1,800 to $2,000, and credit arrangements were not permitted. Almost every bill was paid by the end of the month.

In another case, an electric utility bought some of the water districts (water utilities) in its area. They compared the names on the water bills with the names on the electric bills. Surprise, many of the water bills were in the names of customers who had been disconnected for nonpayment by the electric utility. Again, a substantial amount of revenue was collected.

# Summary

Your utility will never know for sure how much it is losing each year to tampering and diversion, but it does not how much is lost to bad checks, skips, and other forms of fraud. Residential fraud losses far exceed losses to tampering and diversion, although this may not be true with commercial accounts.

A successful revenue protection program must include a strong system for combating fraud losses. This starts by knowing who the customer is before connecting the service, monitoring problem accounts, making it difficult for customers to play the name and revolving door games, and by using skip trace methods to go after customers who leave owing their final bill.

# Recommended Actions

- Review your application for service requirements. Are you obtaining positive identification from both residential and commercial customers? Do you run at least a minimum credit check on new residential customers?

- Do you require a copy of the lease or rental agreement for renters or proof of ownership for new accounts?

- Review your policy for new commercial accounts. Do you require proof of ownership and legal structure?

- Review your bad check policies, particularly for NSF and closed account returned checks. Are your returned check charges the same as the banks in your area?

- When a business goes out of business leaving you with a bill, do you check the legal structure of the business through the Secretary of State or other appropriate office?

- Establish an effective skip trace program. Consider using a credit reporting bureau. Consider using a NCOA database to locate skips.

- When you identify a customer playing the name or revolving door game, investigate other locations where this customer played the same or similar games before confronting the customer. Collect payment for the losses at all the accounts identified.

# References

Federal Bureau of Investigation. Bankruptcy fraud. http://www.fbi.gov/hq/cid/fc/ec/about/about_bf.htm (accessed July 15, 2004).

Federal Reserve System. Check fraud. http://www.federal reserve.gov/publications.htm (accessed July 5, 2004).

Grossman, R. J. 2004. The five-finger bonus. *HR Magazine* (48)10. Also available at http://www.shrm.org/hrmagazine/articles/1003/1003covstory_triangle.asp (accessed July 15, 2004).

National Check Fraud Center. 2003. National Check Fraud Center bad check laws by state. http://www.ckfraud.org/penalties.html (accessed July 15, 2004).

Simmons, M. R. 2003. Recognizing the elements of fraud. In *Articles on Internal Auditing and Fraud Investigation.* http://www.dartmouth.edu/~msimmons/fraud-investigations/fraudwww.htm (accessed July 14, 2004).

United States Postal Service. National change of address. http://www.usps.com/ncsc/services/ncoa.html (accessed July 17, 2004).

# 5
# The Paper Chase and Revenue Recovery

## Sources of Information

The *paper chase* in many revenue protection investigations is limited to the use of utility records. You check the name on the service, consumption and billing records, and the possibility of previous irregularities associated with this customer and/or this location. Occasionally you will have to check with a landlord to determine who is actually renting the property or look up tax records to find out who owns the property. When you are conducting a major investigation, there are a number of additional sources of information that can assist you in conducting the paper chase. Remember the old saying, "no job is completed until the paper work is done."

The investigator's greatest asset is knowledge of sources. The true investigator knows where to go to find the information needed to complete the investigation. A lot of this information is available within the utility, some of it is available from public sources, and occasionally you will need information that is purchased from private sources such as credit bureaus. An experienced investigator will have additional sources of information that are developed over time including Internet resources and individuals with expertise in various areas. There are four general categories of information that you may need to access during the paper chase: government, business, legal and safety, and the Internet. Each of these is discussed separately in the sections that follow.

## Government sources

The government, at all levels, is constantly collecting information on you, me, our homes, our cars, our lifestyles, and other facts. For investigators in the United States there are four publications available from the U.S. Government Printing Office:

- *Where to Write for Birth and Death Records: United States and Outlying Areas* (DHEW Publication Number (PHS) 80-1142)

- *Where to Write for Birth and Death Records of U.S. Citizens Who Were Born or Died Outside of the United States and Birth Certificates for Alien Children Adopted by U.S. Citizens (DHEW Publication Number* (PHS) 80-1143)

- *Where to Write for Marriage Records: United States and Outlying Areas* (DHEW Publication Number (PHS) 80-1144)

- *Where to Write for Divorce Records: United States and Outlying Areas* (DHEW Publication Number (PHS) 80-1145)

These publications are available from:

Superintendent of Documents

U.S. Government Printing Office

Washington, DC 20402

Web site http://bookstore.gpo.gov/

Another important guide is *Braddock's Federal-State-Local Government Directory* (two-volume unabridged edition), which is found in most major libraries. Volume I covers federal agencies and provides specific contact information for organizations and individuals. It cuts through the bureaucracy. Volume II covers sources at state and local governments including specific contact information.

In addition to the library, take a trip to the courthouse near you to learn what information is available there and how to access it. You should also determine what information is available at your state/provincial capital, especially when investigating commercial and industrial accounts.

## Business sources

If you are investigating a major business organization, a trip to the library will provide abundant information. If you are working in the United States, or if the company you are investigating is a U.S. company, the following are good sources of information:

- *The Directory of Corporate Affiliations* (National Register Publishing)
- *Foreign Index to the Directory of Corporate Affiliations* (National Register Publishing)
- *Standard & Poor's: Register of Corporations, Directors and Executives*
- *Thomas Regional Industrial Directory*

For international companies, search *Moody's International Company Data*. This reference contains data on non-U.S. public companies in more than 90 countries.

There are numerous other business information sources available through trade associations and subscription services. Many library systems have business or research librarians who are trained to conduct this type of research, and they have the tools to assist in these investigations. In some cases they will do the research for you, but usually they will show you how to use the tools they have available.

## Legal and safety sources

If you have a law library in your area, plan a visit there as well. The information that will be useful to you depends on the type of investigation you are conducting. You will find copies of the laws for your area, and you may find information on cases involving the organization you are investigating. Among the sources that you will find helpful are

- *Corpus Juris Secundum*
- *Law of Torts*
- Reports of the National Reporter System

Part of an investigator's responsibility is to know how accidents and fatalities can be prevented at revenue protection crime scenes, although the thief is really responsible for the consequences of his or her actions. The meter, seal, and locks, and other venders can be valuable sources of information; however, there are some additional, less obvious sources:

- The *National Bureau of Standards* is involved in building safety, fire research, consumer product safety, and other safety research.

- The *National Technical Information Service* offers more than one million research titles. You can search by topic, review the summaries, and then purchase the documents that are relevant to your investigation.

- The *American Society for Testing and Materials* provides a management system for technical information. Thousands of standards and numerous technical publications have evolved from their committees.

- The *National Fire Protection Association* has experts in fire prevention and investigation.

- The *National Safety Council* (NSC) prepares and disseminates literature, videos, and programs on safety. Of particular interest is the NSC Industrial Data Sheets series.

- *Underwriters' Laboratories* has conducted product testing since 1894 and publishes a catalog, *Standards for Safety*.

To contact these and other organizations that may help you, use the Internet. Go to any major search engine and enter the name of the organization to locate its web site, and you will find much of the information you are seeking. Some of the information needed during a revenue protection investigation may be found in the following sources (these apply specifically to the United States; sources differ in other countries):

- *Civil litigation records:* Legal actions filed at city, county, state, and federal levels provide detailed, documented records of a person's personal history, background, and financial relationships. These records include divorce actions and suits for nonpayment of child support.

- *Criminal histories:* Usually found at the County Clerk's office, these include records of arrests, convictions, and case dispositions.

- *Bankruptcy records:* Personal and business bankruptcies are a matter of public record and show financial relationships and other information on the person or business. These records are maintained at the U.S. Bankruptcy Court in each jurisdiction.

- *Other local government records:* These include property transfer records, marriage licenses, and the like. These records may have the individual's Social Security Number (called *national identify number* in countries other than the United States), which is needed to investigate other sources of information.

- *Corporation and assumed name records:* Corporate filings and records are usually found at the office of the Secretary of State in each U.S. state capital. Local businesses and assumed name records are found at the local courthouse.

- *Uniform commercial code filings:* These filings record the purchases and transactions made by an individual or business and the security interest registered by the parties extending credit.

- *Tax assessor files:* The county recorder or tax assessor maintains public records on everyone in the jurisdiction who pays taxes on real or tangible property.

- *Business information services:* Dun & Bradstreet and Standard Poor's are among the better known services. Many such services are available online.

- *Credit information sources:* Retail credit bureaus provide detailed information on individuals including Social Security Numbers, prior addresses, employers, and credit history.

The following examples give specific information you may need to access in different types of investigations.

## Theft of service investigations
- Property ownership records
- Property rental records
- Customer of record (utility records)
- Financial history
- Employment
- Family connections
- Friends, neighbors, and coworkers

## Fraud investigations
- Civil litigation records
- Criminal histories
- Bankruptcy records
- Corporation or assumed name records
- Credit report
- Previous addresses

## Commercial and industrial cases
- Legal structure of the business
- Names of the owners
- Other businesses owned
- Personal and investment property owned by these people
- The employees at these locations
- Vendors
- Other utility services at these locations

- Criminal and credit histories
- SEC filings and other open-source information on the business and/or the owners

To complete a background check on an individual, you need three pieces of information:

1. Subject's correct name (spelled correctly, determination whether birth or adopted name, determination if person uses nicknames or aliases) may be found on
   - Voting records
   - Name on licenses
   - Credit reports

2. Date of birth may be found on
   - Drivers licenses
   - Some motor vehicle records
   - Professional licenses
   - Fishing or boating licenses
   - Marriage or divorce filing.

3. Social Security Number (or national identification number) may be found on
   - Credit reporting agencies
   - Drivers license in some states
   - Utility records
   - Professional licenses
   - Marriage and divorce filings

## The Internet as an investigative tool

The Internet has become the investigator's best friend. If you know how to search the net, you can locate people, find information on commercial accounts, and search public records. Because the information on the Internet is constantly changing, some of the sources discussed in this chapter may be found at sites other than those suggested here. (And more information and sites are added to the Internet every day.) An investigator must take time to learn to do research on the Internet.

Start with local government web sites. Many of them have property information that includes the name of the owner, assessed value, and property transfers. Some even include photos of properties. Local court records or a list of the records available at the courthouse are available on the Internet. These include criminal and civil litigation records. To find out what is available in your area, go to your favorite search engine and enter the name of the city or county and "courthouse" or "records."

Web sites such as Public Record Finder.com (http://www.publicrecord finder.com), which is available in at least 41 states, can be used to locate birth, marriage, and death records; business licenses; court records; property records; sex offender registries; and other records. It also has links to other sources of information around the world including Canada, Mexico, Europe, and Africa. In addition, Public Record Finder.com provides links to other web sites by category, including "Vital Records Online," "Property Records Online," "People Finder," and "Property Finder." There are several paid features available at this site that will help you locate individuals. This site also features "Net Detective," which has a number of paid features for a reasonable price.

There are a number of other paid background investigation services available on the Internet including the following (prices shown reflect the rate as of July 2004):

- http://www.Intelius.com: Criminal check, legal judgment liens, bankruptcy, address history, and more ($49.95)

- http://yourownprivateeye.com: Comprehensive background checks, online asset search, and criminal record check ($95.00)

- http://www.peoplefinders.com: Current address, 20-year history, criminal background, and more ($39.95)

- http://www.PeopleData.com: 20-year national criminal records search ($60.00)

- http://www.peoplefind.com: 32-point national scan ($69.95)

- http://www.1800ussearch.com: More than 2 billion records, immediate online results (starting at $19.95)

- http://www.ndcr.com: Criminal records in all 50 U.S. states (starting at $39.00)

Two of the most comprehensive paid online services are Dun & Bradstreet and LexisNexis. Dun & Bradstreet (http://www.dnb.com) has specific services for finding people and collecting debt, including a contingent collection service and a Demand Letter Series. These services can be especially useful if the person has moved to another state leaving the utility with a final bill. Dun & Bradstreet can assist the utility in finding credit information and in providing collection activities for residential, commercial, and industrial accounts. LexisNexis (http://www.lexisnexis.com) is more of a research tool offering a wide variety of services that are especially useful to lawyers and law enforcement agencies. These include company information and analysis and links to courthouse records.

There are web sites that conduct an initial locator check at no charge and then provide paid options for additional information. For example, http://www.peopledata.com will tell you if the person is in the database and how many matches they have for that name in the state you search. It also gives you the date of birth. It then gives you the option to purchase locator information, a background check, and a satellite photo of the address listed. The main feature of this web site is that it lists everyone else in the state with the same last name. If you can't locate the customer who owes you money, then you may be able to locate a relative. This is important because people who don't want to be found often put their telephone, utility services, and other locator information in a spouse or child's name. If the person is using an unlisted telephone number, a major source of information for this web site, the name will be listed but the information will not be offered for sale. This tells you they are still around.

Before spending money for the information, however, try some of the free Internet sites. For example, Yahoo (http://phone.people.yahoo.com) will give you the current address if the person's telephone number is listed. If not, it refers you to a fee-based background search service. Other web sites that offer similar search options include http://www.whowhere.com and http://www.smartpages.com. You can also search for businesses on SMARTpages.

Two unique and very effective locators for businesses and people are http://www.switchboard.com and http://kevdb.infospace.com. The results provide the telephone number in addition to the address and a map of the location of the address. The sites also offer additional paid services.

If the customer provided credit card information to the utility, you can conduct a 20-year history check at http://www.peopledata.com/process_form_background.php?. The search looks for bankruptcies, legal judgments, home and property ownership, liens, relatives, roommates, and neighbors for $30.00. They offer an optional national criminal records search for an additional $20.00.

Do you want to spend $50.00 or more to find someone who left you with the final bill? It depends, of course, on amount of the final bill. But in all cases, the free service should be used as a skip trace tool. If the person has moved, it may take some time before the customer's name is back in one of these databases; so if the name doesn't appear immediately, put it in a tickler file and try again in 6, 12, 18, and 24 months. This activity could be assigned to someone in the office who works on the Internet for an hour or two each month trying to locate customers who have left without paying all their bills.

# Computing the Back Bill

When a customer has skipped owing a back bill, that bill may be all you are able to collect. However, if current diversion or meter tampering is involved, additional charges should apply. In addition to paying for the revenues for the utility services that were not billed as a result of the theft, the customer should pay for the cost of the investigation, disconnect and reconnect fees, and, if any are incurred, attorney fees. Some utilities have an established tampering fee that covers the cost of the investigation, which ranges from $100.00 to $500.00. Other utilities bill the customer based on time spent by investigators and other utility employees on the investigation, as well as other direct costs such as vehicle usage. Billing a customer a tampering fee is not an accusation that the customer actually did the tampering, but that the tampering occurred at a location where the customer benefited by receiving services not paid for.

Many utilities also have a cut or damage seal fee that is billed to the customer even if the utility cannot find evidence of tampering or diversion. Customers who report a cut or damaged seal are not billed, nor are they billed if the seal was corroded by weather. The customer is billed only if the cut or damage was obviously done deliberately. Cut seal charges generally range from $25.00 to $100.00.

Fees and charges a customer could incur if meter tampering or current diversion is detected include

- Meter tampering/current diversion charge—$100.00
- Cut seal charge—$50.00
- Disconnect fee—$35.00
- Reconnect fee business hours—$50.00
- Reconnect fee after hours—$80.00
- Meter test fee—$50.00
- Damage to meter—actual costs
- Attorney fees—actual incurred

The fees listed are modest when compared to what some utilities actually charge.

Let's look at an example. Mr. Customer was caught with a hole in his meter. When the utility wasn't around he put a wire through the hole to stop the disk from turning. He forgot to remove the wire on the day the meter was read. The utility believes he has been doing this for several months and computes the back bill for the lost revenues at $500.00. His service has been disconnected. Before it will be reconnected he must pay the bill for charges listed in table 5–1.

Table 5–1 Mr. Customer's back bill.

| Charge or Fee | Amount |
|---|---|
| Meter tampering | 100.00 |
| Estimate back bill | 500.00 |
| Disconnect fee | 35.00 |
| Reconnect fee | 50.00 |
| Meter test fee | 50.00 |
| Total Due | $735.00 |

If Mr. Customer had cut the seal there would be an additional charge of $50.00 for it.

In the United States some state statues allow the utility to collect three times the cost of utility services stolen. In our example, the utility would have back billed the customer $1,500.00 for the energy stolen, and the total bill would be $1,735.00. That gets the customer's attention.

When a customer is disconnected for nonpayment, the utility usually leaves the meter at the location with boots or other such devices to prevent current from passing through the meter. This can be an advantage if the customer decides to remove the boots and steal energy because the utility will have the final reading at the time of the disconnect and the reading when the customer is caught stealing. Assuming the customer let all the energy stolen run through this meter, the utility can determine exactly what the back bill should be. However, computing the back bill is not always this simple, especially if diversion or tampering is found at an active account. Then other methods are used to *guesstimate* the back bill.

If we believe the customer is stealing but there is no evidence at or near the meter, *covert metering* may be used to determine if there is a bypass. A check meter is placed at the transformer, and the readings of the check meter are compared with the reading on the customer's meter. If the check meter shows 1,500 kilowatts and the customer's meter reads 1,000 kilowatts, we know there is a problem. This information could be used to obtain a search warrant to search the premises for a bypass. If the utility doesn't know where the 500 unaccounted for kilowatts are going, there may be an unsafe condition; therefore, the service is disconnected until the situation is corrected. Check meters (i.e., covert meters) are available from several manufacturers, which are listed on the International Utilities Revenue Protection Association (IURPA) web site at http://www.iurpa.org.

Covert metering should only be done at locations where tampering is suspected but the method being used can't be determined, although it is assumed there is a bypass. Some utilities use covert metering at locations where they know tampering is taking place to measure the difference between the actual usage and the metered usage for purposes of computing the back bill. What happens, however, if the utility knows a customer has jumpers behind his or her meter and decides to use a covert meter at the location for a month to compare the readings. During the month the straps short out and cause a fire, resulting considerable damage. The utility knew that an unsafe condition existed but didn't correct it because the bigger concern was computing the lost revenue. Is your utility willing to accept the potential liability in this case?

*Billing records* are often used to compute the back bill. The utility reviews the records to determine when the diversion began and computes the amount of energy it believes the customer actually used based on previous consumption. The difference is the amount to be back billed. Figure 5–1 is an example of an examination of a billing record.

The utility records show when the theft began by the drop in consumption; area **A** on the figure shows the difference between what was metered and the consumption before the theft began. This is the information usually used to compute the back bill. The problem with the calculation is that it assumes the consumption did not continue to increase after the customer started stealing. If anything, the customer's consumption probably increased at a greater rate. There is also a method, called trend analysis, for computing area **B**. Trend analysis uses the known values of a simple linear trend or an exponential growth trend to

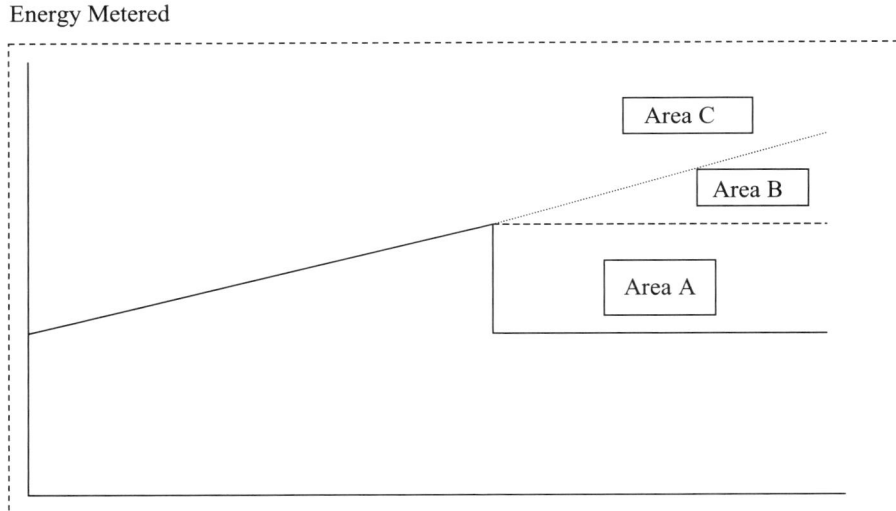

Fig. 5–1. Billing record

project future values. Most spreadsheet programs can project values using trend analysis. On Microsoft Excel® the function is appropriately called TREND. Unfortunately, trend analysis will not compute the unknown dollar amount in area *C*, but if we compute areas *A* and *B*, we are closer to what the customer actually stole.

When using billing records to compute the back bill, go back as far as you can. A smart homeowner may know that reducing the bill more than 10%, will cause his or her account to appear on the utility metering exceptions report. So he or she may reduce the bill 10% for, say, 15 months then reduce it another 10%. After this is done several times, he or she could reduce the original bill by 30% and not appear on the report. And, while the meter readings have been reduced, the consumption has increased. Because he or she controls how much usage is metered, he or she is not worried about running the hot tub constantly or adjusting the thermostat to control his heating or air conditioning.

*Lock and record* is another method used to compute back bills. The customer is caught stealing, but the utility is not sure of the usage so the utility presents the customer with an initial estimate. The utility explains to the customer that it will continue to monitor his or her actual consumption and may present an additional back bill at a later date. For the first month or two the customer will minimize usage. Then, thinking the utility has given up, and because he or she is uncomfortable, usage will go back to normal. The utility should monitor the account for several months after usage returns to normal, then present the customer with the additional back bill.

Some utilities use a *load analysis* to compute the back bill. Using load analysis software, the investigator uses the heating and air normally used for a building of that size and other information to compute the estimated utility consumption. A list of commercially available residential load analysis software is found the Department of Energy (DOE) web site, http://www.eere.energy.gov; however, the easiest way to locate this list is by going to a search engine and entering, "BTS: Building Energy Software Tools." Some of the software listed is free.

Another computational approach to estimating the back bill is a *degree-day analysis,* which compares the outdoor temperature to the standard of 65°F; the more extreme the temperature, the higher the degree-day number. For example, on a day with a mean temperature of 80°F, 15 cooling degree days would be recorded. On a day with a mean temperature of 40°F, 25 heating degree days would be recorded. The degree-day measures can be used to describe the effects of the temperature on the amount of energy needed for space heating or cooling. Degree-day analysis can be computed using an Excel spreadsheet using DDGRAPHS.XLS, which is available free at http://vesma.com/ddd/ddgraphs.htm.

Depending on the method used to steal energy, *testing the meter* may provide the information needed to compute the back bill. If the customer used straps behind the meter, then it could be tested with and without the straps to determine exactly how much energy was diverted through the straps.

A final approach to approximating the back bill is to *compare the usage to a similar account.* If all of the houses in a subdivision are basically the same, the utility can determine the average consumption and how much consumption at this account differed from the average. In the best of circumstances, the utility will estimate the back bill using two of the methods discussed. The second method should help to validate the amount determined by the first method.

# Collecting Lost Revenues

The utility should have a back-billing policy that clarifies the methods used and the amount of time it will back bill, or credit, an account depending on the problems found. If there is a metering or billing problem, this period will be a specific length of time, which may be dictated by state law or utility commission regulations. For meter tampering or diversion, there should be no time limit on the back-bill period. The utility should back bill the customer for the total loss, regardless of the period of time.

When back billing the customer, the utility must provide documentation and supporting evidence including the method used to compute the back bill and the period covered by the bill and the starting and ending dates of the bill. The customer should also be notified that he has the right to dispute the bill.

There are six approaches the utility can use to collect lost revenues.

1. Letter and bill
2. Telephone contact
3. Field interview
4. Interview in office
5. Civil legal proceedings
6. Criminal proceedings

Utilities with a large number of cases or rural utilities where the office may be 50 to 100 miles from the customer may use the letter and bill approach. The customer caught stealing receives a letter from the utility saying that an unusual condition was found at the meter that resulted in a loss of revenue. Because the customer benefited from the condition they own the utility for the loss.

*Dear Customer:*

*On May 14, 2005, a utility employee found an irregular condition on the meter at 14 Johnson Street (the customer's address) that resulted in the failure of the meter to correctly register the amount of* electricity used. Specifically, a hole had been drilled into the meter cover and a wire placed through the hole to prevent the meter disk from turning.

XYZ Utility has reviewed the consumption records at your account for the past five years and determined that the amount of energy that was not metered is valued at $500.00. There are additional charges resulting from the tampering that are listed on the enclosed bill. Since you are the person who benefited from the unmetered energy, according to utility policy and the state statute you are responsible for these charges. Copies of the policy and statute are enclosed.

Payment for the full amount should be received within 15 days to avoid civil action for the collection of the amount on the enclosed bill and possible termination of service.

If you have any questions regarding the meter problem or the bill, please contact me at 999-999-9999.

*Sincerely,*

*Your Friendly Revenue Protection Investigator*

Note that the letter does not accuse the customer of the tampering or diversion. The objective is to collect the lost revenue, not to accuse the customer. The utility knows they did it and they know they did it, but what would an accusation accomplish?

Utilities using the letter and bill approach find that 80% of the letters result in payment in the mail. The other 20% of cases have to be pursued using other means. The problem with this approach is that the investigator does not have the opportunity to let the customers know how serious this crime is and to try, during personal interviews, to get below the tip of the iceberg and find out how the customers learned to steal energy and with whom they have shared this information.

If the customer has been left in the dark, the situation may be resolved by a telephone call, initiated by the customer. The investigator should have all of the information, including the back bill, ready when the customer calls. Again, there is no need to accuse the customer, "some unauthorized person tampered with your meter." The customer won't know who it was and you don't care. You want the lost revenue. If the situation cannot be resolved on the telephone, the customer will have to come to the office.

Some customer revenue protection interviews are conducted in the field at the customer's location. This is usually not the preferred location for a revenue protection interview because the utility employee is on the customer's turf. The employee doesn't know if the customer has a propensity for violence or if there are weapons at the scene. If the interview must be conducted at the customer's location try to have two employees present. In all cases, let someone at the office know where you are and why and have a specific call-back time. If you don't call back by that time, the police need to come looking for you. The preferred location for a customer interview is at the utility office where the customer is on your turf and you control the dynamics of the interview. Specific approaches and consideration for the revenue recovery interview are discussed in the next chapter. Chapter 9 discusses the civil and criminal approaches to revenue recovery. As we have already discussed, in some cases the utility may use both of the legal options.

## Summary

Most revenue protection cases are simple; you disconnect the customer for nonpayment and they reconnect themselves. You catch them and collect revenue for the services used since the disconnect. But the true investigator is prepared to investigate the more complex cases; residential theft by active customers and commercial theft. These cases may require the investigator to conduct a paper chase to get below the tip of the iceberg. The investigator should know what information is available at the local library and courthouse and how to "let their fingers do the walking" on the Internet.

Because one of the primary objectives of the revenue protection program is to collect lost revenue, the utility must understand the different approaches to computing the back bill. It must also decide which methods it will use to collect those lost revenues.

## Recommended Actions

- Review the list of information resources and decide which ones you should order. Visit the reference section of the local library and visit the courthouse. Identify the information at these resources and learn how it is obtained.

- Become familiar with the information resources on the Internet. Run a search for information on yourself to test the resources listed. Are there paid Internet resources that would assist you in your investigations?

- Review the list of revenue protection–related charges and fees. Also check with other utilities in your area to compare charges and fees. Do you need to make any changes?

- Determine which methods you will use to compute back bills. Review the software discussed to determine if any of these programs would assist you in computing back bills.

- Consider using trend analysis when using consumption records to compute back bills.

- Develop a back billing policy and identify which of the six approaches listed will be used to collect back bills in revenue recovery cases.

# The Investigative Interview

## Planning for the Interview

In general, interviewing is considered an information-gathering exercise. When conducting a revenue recovery interview, however, the primary goal is the recovery of lost revenues. In most cases, it is not really important to get the customer to confess that he tampered with the meter. What is important is that he admits he used the utility services that were not paid for and that he owes the utility for the loss. Remember, the number one goal in most interviews is the recovery of lost revenues.

There are two primary types of interviews, standardized and nonstandardized. A standardized interview is structured, and includes an interview form with questions that are asked at each revenue protection interview. A number of utilities use this approach. Its advantage is that you can ensure that the questions asked address the elements of the crime needed to prosecute in your jurisdiction. There are also obvious advantages when working with newly assigned investigators, and it becomes an important element in their on-the-job training.

The two types of standardized interviews are scheduled interviews and nonscheduled interviews. In a scheduled interview the suspects are asked the same questions, in the same order, during each interview. During a nonscheduled interview the interviewer may change the order of the questions and he or she may use alternative wording.

The elements of a standardized interview usually include the following:

- The content of the questions
- The exact wording of the questions
- The context to be supplied with each question
- The sequence of the questions to be asked
- The answer categories, if any, into which the responses can be placed

The standardized interview is not inflexible. The interviewer may adjust any of the components within predefined parameters. Another aspect of the standardized review is topic control. This is the extent to which the interviewer controls the topic of discussion and takes the initiative to direct the course of the interview. To maintain topic control you must first define the central focus of the interview and the boundaries within which you will operate. If the subject tries to wander off into topic areas that are not relevant to the interview, the interviewer uses the predetermined questions to return the focus to the objective, revenue recovery.

A nonstandardized interview approach gives the interviewer some freedom. The interviewer can select questions specific to a case and word them according to his or her own objectives and style. Although standardized interviews are appropriate for newer investigators, experienced investigators usually prefer the nonstandardized approach.

A balance between standardized and nonstandardized interviewing can be achieved using an interview guide that provides an outline or checklist of the topics and subtopics to be covered during the interview, but does not provide a specific sequence of questions. The interview guide has two functions. First, it provides a reminder to the interviewer of the areas to be covered during the investigation. Second, it helps the interviewer maintain an inventory of what has been covered and what has not.

Interviews do not take place in a vacuum. There are three primary dynamics to the interview: the interviewer, the respondent, and the question. There is also consideration of the culture of the community, the interviewer, and the respondent. As in all successful communications, it is the responsibility of the communicator (i.e., the interviewer) to communicate according to the needs of the person he or she is communicating with (i.e., the respondent or customer). Some factors regarding the primary dynamics follow:

1. *The Interviewer.* An interviewer must develop a personal style based on his or her personality. What works for one person, may not work for others. As you read about and practice different interview techniques and approaches, adjust each so that it fits your personality.

2. *The Respondent.* The interview should be adjusted to fit the interviewee's culture and personality. You must be able to communicate at a number of different levels and with a number of different types of customers.

3. *The Question.* Each question asked should have a specific objective. It should be to the point and easily understood by the respondent.

The interviewer must not contaminate the information obtained during the interview. Contamination occurs when the investigator impedes or negatively influences the interview process, causing the respondent to provide inaccurate information. One of the factors that can result in interview contamination is the environment in which the interview takes place. The location should be free of distractions, which is one of the reasons that field interviews are challenging. During a field interview the respondent may be distracted by a passing vehicle or wondering if the neighbors are watching him or her being questioned by a utility employee while standing at the meter.

Interview contamination can take place if there are too many interviewers present. Unless there are highly unusual circumstances, no more than two interviewers should be present. If possible, only one interviewer should be present in the room when conducting the interview at the utility office. According to John Reid, a renowned interviewer and polygraph operator, "The principal psychological factor contributing to success…is privacy." By conducting the interview at the utility with one interviewer you have maximized this principle.

Experienced interviewers know that what the respondent says verbally may be less important that what is said nonverbally. It is important that the interviewer be equally aware of his or her nonverbal behavior. Use an open and relaxed posture, facing the respondent; lean forward and maintain eye contact, nod, and occasionally say "uh huh" and "okay."

Interviewers must not only be aware of their verbal behavior (the language they use) but also their *paralanguage*. Paralanguage is the manner in which a person speaks (such as tone of voice) and can communicate meaning beyond the words spoken. The best interviewers slow down their rate of speech and speak softly. This may seem unnatural at first but it will elicit more information from the respondent than speaking louder and in a normal or fast rate of speech. When asking questions, do not place more emphasis on any one word over any other. This is referred to as phrasing questions in a "leveler mode." Some interviews are contaminated because the questions are asked in a random or haphazard manner. To prevent this from occurring, the interviewer might consider asking questions in the chronological order in which the events occurred. "When did you move to this address?" "When did you first notice a change in your utility bill?" "What did you think caused the reduction in your monthly bill?"

To maximize the effectiveness of the interview while minimizing contamination, consider using the funnel approach. This begins with the broad end of the funnel using open-ended questions and work toward the narrow end of the funnel and closed questions. Open-ended questions require the respondent to reply with a narrative. "Tell me what you think happened to your meter?" Closed questions require a specific response. "Did you personally tamper with this meter?" As illustrated in figure 6–1, the funnel approach goes from the general to the specific.

Interview research over the past 70 years indicates that the use of open-ended questions generates more complete information but that the information may be less accurate than responses elicited by closed questions. But by starting with open-ended questions the interviewer can assess the respondent's behavioral norms. How does the customer react to different questions in terms of the verbal response, paralanguage, and nonverbal behavior? If there are changes in these responses during closed questioning, the interviewer is better prepared to recognize the changes taking place. Verification questions are used once the respondent has presented his or her *story*. Assessment questions are used when the interviewer detects possible deception. These are discussed later in this chapter.

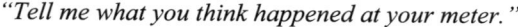

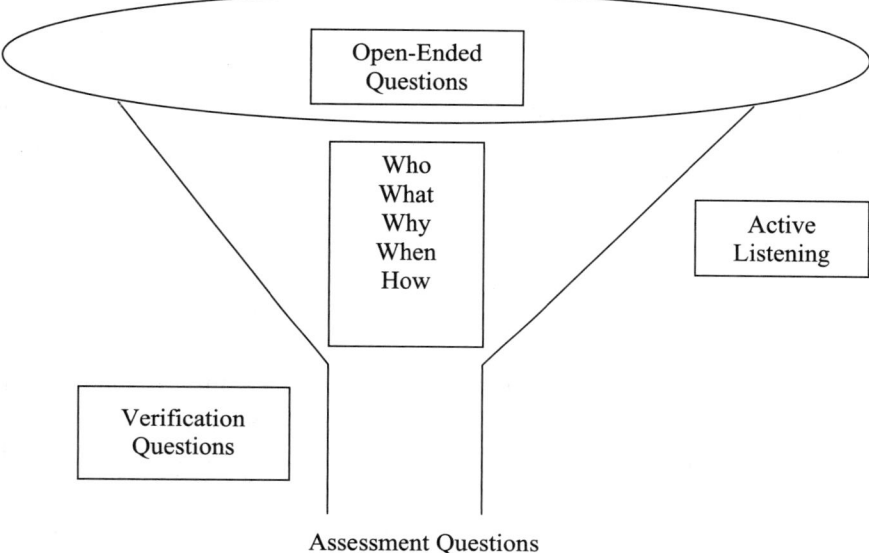

Fig. 6–1. The funnel approach to interviewing

There are four main elements to any interview process:

1. Purpose of the interview

2. Information that has already been gathered

3. Information that is to be acquired

4. Strategy used to obtain the information

The primary purpose of a revenue recovery interview is REVENUE RECOVERY! Even if you want to question the customer about the methods used or the possible involvement of a fixer, go for the revenue recovery first. If the back bill is not paid in full during the interview, ask the customer to sign a promissory note agreeing to the amount owed and the terms agreed to, and then begin asking questions about the fixer or the method used. The process should be thought of as two separate interviews, one (which is the first interview) for revenue recovery, and, once that is accomplished, the second interview may be used to try to identify the fixer.

You should have as much information on the customer and the account as possible before scheduling the interview. This includes consumption history, previous locations in your utility service area, and real estate tax records or a copy of the rental agreement. You also should have the information and evidence gathered during the crime scene investigation. But don't attempt to impress the customer/thief with your thoroughness. Think about holding back some of the information and evidence to see if the customer contradicts some of the things that you already know. As the saying goes, "Don't put all of your cards on the table."

Because the primary objective of the interview is revenue recovery, the information you will want to obtain is anything that will substantiate the back bill. Then, as we have discussed, you may attempt to get additional information including the name of a fixer or the source from which the customer learned how to steal. Information to obtain may include the date the customer moved to the location, where else they have lived on your system, and any major changes in lifestyle or appliances that may have impacted power usage. In most cases your interview strategy is simple; conduct it as you would any other formal business meeting. Do not play games, like "good cop-bad cop." And do not attempt to use any of the advanced interview approaches discussed in the final section of this lesson unless you are thoroughly trained and experienced.

There are at least two other interview constraints. The interview must be conducted in an ethical manner. It has to be fair and should not be oppressive. In addition, you must be aware of the evidential criteria resulting from the interview. The information you obtain should be admissible if you do go to court, and it must be able to withstand the tests of reliability and validity. Reliability means you would get the same answers if you asked others (with the same knowledge as this customer) the same questions. Validity means that the information is correct and is usually substantiated with additional evidence.

# The Interview

The revenue recovery interview will take place either in the field, at the customer's location, or in the office. On rare occasions the interview will take place at a third-party office, for example, the prosecuting attorney's office.

## Field interviews

In most cases, it is preferable to conduct the interview at the utility where the interviewer has greater control over the environment. Occasionally a revenue recovery interview is conducted in the field, at the customer's location. When this occurs, consider the following suggestions:

- If possible, conduct the interview outside. Once you enter the dwelling you are on the customer's turf where he or she is in control. Besides, the customer may have weapons inside of the house and be prone to use them.

- More than one employee should be present. If this is a repeat offender or if you are in a dangerous location, consider asking a police officer to escort you during the field investigation. Conducting the interview alone and going inside the customer's house violates two basic rules, and anything could happen.

- If you have to conduct the interview alone, let someone know where you are and have a specific future contact time. If you do not report back by that time, the police should be notified immediately.

- Use a voice-activated mini tape recorder to record the conversation. Don't hide it on your person; hold it in your hand.

- Even though you are on the customer's property, be firm and in control. However, do not be intimidating.

- Have an escape route planned, and if the customer becomes aggressive, use it!

## Office interviews

At the utility's office, the interviewer is in a much better position to control the dynamics of the interview, and the customer is less likely to become belligerent or aggressive. To set the scene for the office interview, consider the following:

- The customer(s) should sit closest to the door. There have been law suits filed and lost because an investigator (not a law enforcement officer) sat between the interviewee and the door, implying unlawful detention. During a revenue protection interview, the customer is free to leave at any time.

- Conduct the interview as a business meeting, preferably across a table or desk.

- Openly tape record the interview. Tell the customer this is for the record. Do not use a hidden tape recorder. In many jurisdictions this can lead to serious legal complications.

- The room should say "gotcha" without being threatening. You might have the case file on the desk, photographs, and possibly some of the evidence, to show the customer that you are prepared. You will also have a back bill and expect to be paid. Remember it is your manner and attitude that is most important. Be professional and courteous. Not threatening.

- Do not accuse the customer directly of the tampering or diversion. "Some unauthorized person tampered with the service and as a result the meter was not operating correctly." There are cases where you do want to accuse the customer to gauge their reaction. But make sure that you have a reason before doing so.

- If the customer arrives with an attorney, the interview is over unless your attorney is also present. Attorneys talk to attorneys. The rest of us mere mortals talk to each other. If the customer's attorney is present and yours is not, you may say something that is better left unsaid.

The interview may include three stages, especially if the customer is questioning the amount of the back bill or responsibility for the loss.

The first stage is the *general questioning* stage. You ask the customer questions and he or she provides answers. Try to use open-ended questions as much as possible. The *examination of the information* is the second stage. This is where you confirm the information obtained during the first phase. Using the funnel approach, this is what is referred to as verification questions. You reconfirm dates and other information provided by the customer. The purpose of stage two is to attempt to confirm the information that is true and identify possible deception. The final stage is the *cross-examination*. If you identified possible deception during the second stage, this is where you challenge the customer. Don't say. "You are lying to me." Point out the inconsistencies and ask the customer to explain. Maintain your professionalism even when lied to.

The reason you should use a tape recorder during an interview is so that you can concentrate on the customer's nonverbal behaviors and record changes in demeanor when certain questions are discussed. If the customer is giving you a story you do not believe, use the three-stage approach and have him or her repeat the story. If the person can't give you the same story the second or third time that was told initially, you *got' em*.

When conducting the revenue protection interview, note everything that is said and how it is said. This is why you use the tape recorder. Also document in the notes of the meeting the respondent's demeanor when responding to your questions.

Whether you are interviewing in the field or in the office, use open-ended informational questions at the beginning of the interview. Save the closed questions for the later part of the interview and avoid leading questions unless you have a specific reason for asking them. In many cases leading questions result in interview contamination. Informational questions are nonconfrontational, nonthreatening, unbiased, and designed to gather information. When using informational questions, consider the following:

- They are not likely to cause the interviewee to become defensive or hostile.

- Facts are obtained in order of occurrence or some other systematic order.

- Only one question is asked at a time.
- Questions are straightforward and frank.
- The respondent is not rushed.
- The interviewer does not suggest answers.
- The interviewer is careful of nonverbal behaviors.
- Questions may be repeated or rephrased if necessary (because they are not understood or the initial answer is not satisfactory).
- The interviewer must ensure that he or she understands the answers.
- The customer must have the opportunity to qualify answers.
- Facts should be separated from inferences.
- The respondent might be required to give comparisons by percentages, fractions, estimates of time, or other measures.
- The interviewer must ask appropriate questions to get all the facts needed to achieve your objectives for this interview.

Let's assume that your goal is not just revenue recovery. You have a customer whose meter has been altered by a professional fixer. It is the tenth time you have seen this method used and the other nine people aren't talking. Your primary objective has changed. You are not just conducting a revenue protection interview; you are conducting an investigative interview. Your first goal is to make the customer feel at ease. To make the interview as pleasant an experience as possible for him or her. You accomplish this by using a friendly (yet professional) and considerate approach. You don't want to become the customer's best friend, but you do want to appear to be an "understanding" investigator who has genuine concern for the customer.

The second goal is to create an attitude of willingness to be interviewed and willingness to cooperate; in other words, a willingness to tell you everything the he or she knows about the matter being investigated. The third goal is to cover all of the issues and elements of the case to determine what the customer knows about each point. This is accomplished by using a list of prepared questions. The list should cover the basics: what, where, when, who, how, and why. The final goal is to make the customer willing to sign a summarized statement of the facts revealed during the interview. This statement may be needed later in court when the interviewee denies making the statements provided during the interview.

With these four objectives in mind, consider in chronological order the six functional steps of an investigative interview:

1. Create a favorable atmosphere for the interview by establishing a rapport with the customer.

2. Allow the customer to provide a free narrative of the information relating to the case.

3. Clarify the information obtained during the free narrative during the question and answer stage of the interview.

4. Verify these facts by asking additional questions to clarify the information obtained and to identify possible deception.

5. Summarize all of the information obtained during the interview.

6. Close professionally. Thank the customer for his or her cooperation and leave the door open for future contact. You may want to ask the customer to sign a transcript of the interview or at least a summary of what has been discussed.

There are a number of reasons why the respondent will withhold information during the interview:

- Protection for self, family, friends, or the suspect's business.

- Fear of becoming involved, of personal retaliation, or threat of harm to the suspect or to people around him or her.

- Personal involvement—the suspect does not want to admit involvement or otherwise jeopardize him or her self.

- Nature of the subject—too risky or too unpleasant to discuss.

- Personal risk or inconvenience. This usually applies to witnesses rather than the suspect. The respondent doesn't want to get involved.

- Failure to ask for the information. If the interviewer doesn't ask the right questions, the interviewee is not under any obligation to provide the information.

# Detecting Deception

Believe it or not, on rare occasions the people you are interviewing during a revenue recovery interview will lie to you. How do you detect deception and how do you respond? Once again the first consideration is revenue recovery. Do you really care if they lie to you as long as they agree to pay the back bill? But, it's always nice to be able to tell if the interviewee is being deceptive.

There are three types of lies: concealing lies, falsifying lies, and definitional lies. A *concealing lie* occurs when the interviewer does not ask the right question. The customer knows that you want to know who altered the meter but you don't ask. This is the easiest lie to commit, especially for the honest thief. A *falsifying lie* occurs when you ask the question and the customer has to say he or she does not know the correct answer. You ask, "Who might have altered your meter?" The customer says, "I don't know," when in fact he or she does know. This is a more difficult lie for the honest thief, but not a problem for the dishonest thief who lies all the time. A *definitional lie* usually is the result of not defining the question correctly. You ask, "Did you do this to your meter?" Rather than asking, "Do you know who did this to your meter?" The customer didn't do the tampering, but knows who did. The response is directly tied to the question, not to what the interviewee knows the interviewer wants to know.

Here are some verbal clues that may indicate the person you are interviewing is lying to you:

- *Changes speech patterns:* If you expect a difficult interview, have a friendly chat with the customer before you begin the formal interview. "How is the weather?" "Do you like living in our community?" "Do you have any relatives on the system?" You are establishing a baseline from which to evaluate if the customer deviates during the interview. Then look for deviations from this baseline during the interview. If the speech pattern changes, this may be an indication of stress. The stress may be related to deception, or it may be because you have caught an honest thief stealing.

- *Repeats the question:* You ask, "Do you know who did this to your meter?" The customer responds, "Do I know who did this to my meter?" The interviewee is stalling for time while he or she thinks up a response. The normal human speech pattern is 120 to 140 words per minute. But we think at up to 450 words per minute. We use our speech to stall while our mind thinks of an excuse.

- *Makes comments regarding the interview or the environment:* "Gee, isn't it hot in here?" or other comments about the interview or the environment are a way of stalling while the interviewee thinks of a lie to answer your question.

- *Exhibits selective memory:* The interviewee remembers events that help to prove his or her innocence or noninvolvement, but does not remember events that may suggest guilt.

- *Makes excuses:* The interviewee says, "I'm a little nervous," or similar comments that indicate he or she is uncomfortable in this situation. Not because he or she is guilty, but because he or she is not accustomed to being accused of a crime.

- *Uses oaths:* If the interviewee says, "Honest, I won't lie to you," think, LIE! This is a standard indicator of deception.

- *Gives character testimony:* The interviewee doesn't expect you to take his word that he is honest, so he invokes someone else's testimony. "Check with my boss, I would never do something like this." "My wife will tell you, I didn't know this was happening."

- *Answers a question with a different question:* You ask, "Do you know who might have done this to your meter?" The interviewee responds, "My meter, let me think, who might have done this…?" Again, the customer is stalling for time while he or she thinks of a deceptive answer.

- *Overuses respectful remarks:* If constantly saying "Yes sir" (or "Yes ma'am)," the interviewee is uncomfortable and trying to impress you with exaggerated respect for your position.

- *Gives increasingly weaker denials:* This results from using the three-stage approach of interviewing. The interviewee is definite during the initial stage, but becomes increasingly weaker as you progress though the second and third stages of the interview.

- *Says more than "No" when denying knowledge of fact:* This is when the interviewee is elaborating. When no elaboration is required, he or she is again stalling for time.

There are also nonverbal cues that indicate possible deception:

- *Full body motions:* The interviewee changes posture when answering questions.

- *Anatomical physical responses:* These are involuntary responses that occur when someone is being deceptive. Without getting into the role of the autonomic nervous system, these responses are the result of the mind-body interaction. When the mind lies, the body sends out signals. This may be grimacing, blinking of the eyes, or other cues. Watch for changes in these responses during the interview.

- *Illustrations:* The interviewee makes motions with the hands to demonstrate points while talking. In some cultures, hand motions accompany all verbal communications. But look for increases in these motions during the interview.

- *Hands over mouth.* This is an indication of a subconscious reaction to concealing a statement. The interviewee knows he or she should say something but is withholding the information.

- *Manipulators:* Actions like picking lint from clothing, playing with pencils, and the like. may be a sign of nervousness, which could be an indication of lying or simply an indication of nervousness regarding the interview.

- *Fleeing position:* (Remember that the interviewee sits closest to the door.) This signal occurs when the interviewee is facing you but his or her body turns toward the door, indicating that subconsciously he or she wants to be somewhere else. The head is answering your questions but the body is ready to leave.
- *Crossing arms:* This is the classic defensive reaction. But don't be too quick to make a judgment in response to crossed arms. It may simply be a comfortable position for the interviewee or maybe it's cold in the interview room.

Although each of the above could be an indicator of deception, look for multiple indicators, both verbal and nonverbal. And go with your gut-level feeling. After you have been working in this field for awhile, you will know when you are being lied to. Listen to your inner feelings. If you think the customer is lying, you are probably right.

Another technique for detecting deception is statement analysis. Statement analysis is the process of examining a person's words to determine exactly what is being said. This includes determining if the person is being truthful or deceptive, discovering additional information from within the statement, and determining if the person is withholding information. To a great extent statement analysis techniques are based on word definitions. Every word has a meaning, and people usually mean exactly what they are saying. Consider the statement, "You know, I am trying to be as honest as possible." The respondent starts off saying, "You know," expecting you to take for granted that he or she is being honest. However, he or she did not say, "I am being honest." We can't believe the interviewee is being honest until he or she tells us so, and maybe not even then.

The person goes on to say, "I am trying to be as honest as possible." The word "trying" means he or she attempted to be honest but failed. The word "possible" means there is a limitation to his or her honesty.

Statement analysis is based on the different parts of speech. To analyze a statement, the investigator first needs to examine these parts, particularly pronouns, nouns, and verbs.

## Pronouns

Common examples of pronouns are I, me, you, he, she, we, they, and it. In statement analysis, particular attention should be given to the personal pronouns "I" and "we" and all possessive pronouns such as my, our, your, his, and her. Investigators have noted that truthful people give statements using the pronoun "I," which is the first person, singular. If the respondent starts answering questions using the pronoun "I" but then later uses "we," the investigator should question why. A change in the use of pronouns may indicate the person is trying to remove himself or herself from personal involvement. If the respondent uses the pronoun "we" at the beginning of the interview it may indicate that more than one person is involved in the tampering. If the person starts with "we" and then refers to "they," he or she may be trying to distance himself or herself from the others.

Watch for the use of possessive pronouns (my, our, your, his, hers, and their), which may reveal the attachment of the person to another person or object. If the customer begins by referring to my or our house and later uses the word "it" when referring to the house, he or she may be attempting to distance himself or herself from the tampering.

## Nouns

Nouns denote persons, places, and things. But they have different meanings for different individuals. Again, the investigator is looking for a change in the use of nouns during the interview. If the customer repeatedly refers to his brother, who you believe was the fixer, as "Bill," then why later in the interview does he or she refer to him as my brother? It may be the customer wants to subconsciously distance himself or herself from Bill by using a less personal noun.

## Verbs

Verbs express action, either past, present, or future. In statement analysis the tense of the verb is important. In a truthful statement the use of the past tense is the norm because by the time the person relates what happened it has already occurred. If the customer uses the past tense at the beginning of the interview and then changes to the present tense, the investigator should be suspicious.

## Repeated accounts

Another aspect of statement analysis is comparing repeated accounts of the same events. Truthful accounts may have a slight variation because the person is depending on memory. A deceptive account must be retold carefully to avoid any discrepancies with the information previously provided. Statement analysis is an emerging scientific approach to detecting deception. Although it is applied mostly to written statements, it can also be used during interviews. Because you are recording your interviews, listen to the tapes when riding in your vehicle. Listen for statement analysis indicators of deception. This is also a good opportunity to critique yourself and to improve your interviewing skills.

In the funnel model we mentioned assessment questions. These are asked toward the end of the interview if you believe the respondent is being deceptive. This is the cross-examination phase of the interview. Assessment questions are closed questions that confirm information the respondent has already given. "You said you were on vacation deep sea fishing when the meter was tampered with. But I checked the weather and there were high seas and storms all that week. Did you get seasick?"

Remember that your primary objective is to recover revenues. If you are going into the cross-examination phase using assessment questions, get the money first. Assessment questions are used if you think the customer hired a fixer and you want to know who that is, or if the customer may know of others who are stealing services.

# Advanced Interview Techniques

So you think you are ready to become a highly professional criminal interrogator and interviewer. One of the best courses to help you achieve that goal is offered by John E. Reid and Associates, the Reid technique. There is a schedule of courses on their Internet web site. If there is no course scheduled in your area, order the book, *Criminal Interrogation and Confessions*. Part of the Reid technique is the following nine steps to interrogation.

Step 1 is a direct, positively presented confrontation with the customer. This is a statement that as far as you are concerned this is the person who is responsible for tampering. Before going to the next step, repeat the accusation with even greater conviction than the first time.

Step 2 is where you, as the interrogator, express a supposition about the reason for the crime's commission. Offer the suspect a possible moral excuse for having committed the crime. Fix the reason for the crime on someone else or on the unique circumstances the subject was in at the time the decision to tamper was reached.

During step 3, you cut off the denials of guilt. Return to the moral excuse you proposed in step 2.

Step 4 involves the task of overcoming the suspect's comments about the moral excuse presented in step 2. These continued denials usually attempt to explain how or why the suspect could not have committed the crime.

Step 5 is where you elicit and maintain the suspect's full attention. Be sincere in your efforts to help the person and demonstrate this sincerity in your verbal and nonverbal communications.

Step 6 is used if the suspect becomes quiet or withdrawn. It's time to reinforce your sincerity by maintaining good eye contact with the customer. The change in demeanor may be a sign that the customer is giving up.

Step 7 is the use of an alternative question, a suggestion of a choice to be made by the customer between an acceptable and an unacceptable aspect of the crime. The question could be, "Was this the first place you have lived where you have bypassed the meter?"

During step 8 the suspect tells you the various details of the offense. When did the tampering start? Did someone show you how to do this? You want the customer to confess to all of the elements of the crime needed to obtain a conviction if the case goes to court.

Step 9 is when you have the suspect put the confession into writing, sign it, initial the bottom or top of each page, and have the signature witnessed.

There are three important points to think about before you consider using the nine-step approach.

1. This approach is used only on very special revenue protection investigations. In most interviews you are not concerned with a confession; your objective is to recover the lost revenue. The Reid approach is appropriate when you are trying to identify a fixer or if you are working with law enforcement because you suspect that this person is responsible for other crimes as well.

2. Not every interrogation will include all nine steps, and the steps do not have to be done in the order they are presented. The experienced interrogator will know which steps to use and in what order to use them.

3. At each step, be prepared to evaluate the behavioral responses the suspect is displaying. These include the verbal and nonverbal responses discussed previously.

There are two characteristics that professional interrogators must continually practice. The first is *patience*. Confessing is not a natural thing to do, and we should expect hesitation and reluctance on the part of the suspect. When you and the suspect feel that the interview is over, keep going. Act like you have all the time in the world and act as though this is your most important case at this time.

The other characteristic is *persistence*. Again, when the customer thinks the interview is over, go back to a discussion of the main aspects of the case. Don't start talking about irrelevant subjects. Give it another 15 minutes.

The customer is about to confess but asks, "If I tell you about this, what can you do for me?" Make no promises! A promise could jeopardize any legal actions that result from the investigation. A customer who asks that question is telling you that he or she is ready to confess. Why make any promises?

# Summary

The only way to become an effective interviewer is to practice, practice, practice. During the early part of your career you should use a structured interview approach. Develop a set of basic questions and review them with your supervisor. As you become more experienced, you will rely less on the structured interview and adjust each interview to the particular case you are investigating.

A final suggestion when conducting a revenue recovery interview (assuming you did not ask about fixers) is that when you complete the interview and the customer is ready to leave, ask, "Who else's meter do I need to look at?" You have caught the customer off guard. The response you get may surprise you.

# Recommended Actions

Develop a basic set of questions for a revenue recovery interview. Assume that you are interviewing a residential customer who had an unauthorized meter in the socket. Your objectives include the following:

- Collect lost revenues and investigation costs.

- Learn if the customer had any changes in load factors (use of your service) during the period of the theft.

- Determine if the customer did the tampering or whether a fixer was involved.

- Determine if the customer knows anyone else who is stealing from your utility.

- Define other objectives you feel are relevant to the situation.

Consider participating in a professional investigative interviewing seminar or course. Check with your local law enforcement agency for recommendations.

# References

Adams, Susan H. Federal Bureau of Investigation. 1996. Statement analysis: What do suspects' words really reveal? http://www.fbi.gov/publications/leb/1996/oct964.txt (accessed July 22, 2004).

Inbau, F., J. E. Reid, and J. P Buckley. 1986. *Criminal Interrogation and Confessions,* 3rd ed. Baltimore: Williams and Williams.

McClish, M. Pronouns. 2003. http://www.statementanalysis.com/pronouns (accessed July 22, 2004).

Sandoval, V. A. Federal Bureau of Investigation. 2003. Strategies to avoid interview contamination. *FBI Law Enforcement Bulletin* 72 (10):1–11.

# 7
# Investigations at Potentially Dangerous Accounts

## The Problem

The customer, who lived on a small farm, had been disconnected for nonpayment of a bill he disputed. He reconnected his service several times and so a lineman was sent to disconnect the service at the pole. The lineman had climbed the pole and was about to disconnect the service when he looked down and saw the customer pointing a shotgun at him. "You're not going to disconnect that service," he was told. The lineman agreed. The lineman slowly descended the pole as the farmer, who continued to point the gun at him, yelled at him that the utility had made a mistake on his bill. The lineman agreed that there had obviously been a mistake and said that he would go to the office and look into the problem. He then asked the customer's permission to go to his truck, removed his gear, and slowly drove off the property. Several hours later sheriff's deputies arrived and arrested the customer.

In another case the meter reader didn't realize the building was a crack house. All he knew was that someone had put a meter into the meter base and was stealing electricity at what was supposed to be an inactive account. Before he could report the theft to the utility, the "crack heads" spotted him and thought he was calling the police. They forced the meter reader into the house, tied him to a chair, packed up their gear, and left. He wasn't found until later that evening by the police who began searching for him when he didn't return to the utility at the end of the workday.

A revenue protection investigator was at a residential service where tampering had been reported. The meter was under a second floor deck. As he was conducting the crime scene investigation, he had a feeling that someone was watching him. He looked up to see the customer on the deck above him with a .22 caliber rifle. The customer leaned over the rail of the deck and emptied the rife at the investigator as he was running from the scene. Fortunately, the customer was a lousy marksman and the investigator was not injured. The customer was sentenced to six months in jail and 30 days of community service.

As an African American meter reader was reading his regular route, he came to a house where the meter was lying on the ground and jumpers were in the meter socket. Suddenly a huge white man came from around the corner of the home and told the meter reader, in very colorful terms, to get off his property. The meter reader immediately returned to the utility and reported the incident. His manager called the local police and learned that the customer was the head of the local Ku Klux Klan Klavern. He also learned that the customer had been in prison twice for homicide and that he was currently on parole. As a result of this incident, he was returned to prison to finish his sentence for his second murder.

Revenue protection investigations would be easy if all the people caught were honest thieves and if most of them had been disconnected for non-payment and had reconnected themselves. But it's not that easy. Some of the people you investigate will be engaged in other criminal activities and some of them are just plain mean.

Theft of utility services at drug manufacturing and indoor marijuana grow operations is a problem around the word. While the *Atlanta Journal-Constitution* (June 6, 2004, pp. C1–C4) was printing a story about the surge of methamphetamine (meth) labs in both urban and rural Georgia, police in Fiji were raiding a huge meth lab that had enough chemicals to manufacture $540 million worth of the drug. According to the article on meth labs in Georgia, seizures in 2003 in rural areas reached 80 per 1,000 population, compared to 30 seizures per 1,000 population in 2000. In urban areas, there were 20 seizures per 1,000 population in 2000 and 40 seizures per 1,000 population in 2003. The meth lab on Fiji was operated by an Asian organized crime group that was exporting the drug to Australian, New Zealand, European, and American markets.

Marijuana growers have to worry that the neighbors or the meter reader will spot the plants if they plant them in the yard so they have moved indoors. Indoor marijuana growing has become a major industry, and many of the growers are stealing energy. An indoor grower may use 10 times the amount of energy the building would normally use at the location. Two growers in New York stole approximately $70,000 of electricity over a six-month period. Growers in California report on the Internet that power is their biggest cost. In Florida, outdoor growers complained about the drought and moved their plants indoors.

These cases are but a few of many such true stories, and they illustrate that revenue protection investigators must constantly be alert to potential dangers in the field. Other field employees, including linemen and meter readers, should have the same awareness and know how to recognize and respond to potentially dangerous situations.

## Drug Manufacturers

Drug dealers and manufacturers may be producing or selling any number of illegal drugs including MDMA (Ecstasy), GHB (sometimes called Liquid Ecstasy), ketamine, cocaine, crack cocaine, and methamphetamines. Some of these drugs, such as GHB and ketamine, are smuggled into the United States and other countries where they are illegal. But large quantities are also manufactured locally.

Smaller drug laboratories may not be stealing utility energy, but they still pose a danger to utility employees. Many of these labs are set up in the manufacturer's home or apartment and don't use excessive amounts of energy. But many of the manufacturer's are also users who are paranoid, and they may react violently if they think a utility employee found out what they are doing. Meth labs, as we discuss later, present a special danger because the chemicals they use are carelessly discarded, and an employee may come into contact with them and not know it.

MDMA, or Ecstasy, is a "designer drug" that was first synthesized in 1912 as a precursor agent (i.e., an intermediate structural compound, possessing properties that contain primary constituents for therapeutic compounds). It induces a five- to seven-hour euphoric effect characterized by increased activity, mood alteration, and altered perception. The pills or capsules cost $15 to $25 a hit in most areas, $35 to $45 in other areas. Although it is supposed to be the "love drug," MDMA actually tends to decrease sexual behavior. There are also a number of cases where MDMA use has resulted in death, especially if the users take the drug with other substances such as alcohol or cocaine. The drug is popular among some of the college crowd and is used frequently at rave parties. Although 90% of the Ecstasy in the United States comes from the Netherlands and Belgium, where it is also illegal, the other 10% is manufactured locally. This is a problem because the manufacture of the drug is extremely hazardous, and the chemicals used are hard to obtain legally. One police officer involved in a recent arrest described the Ecstasy cooks as "self-taught." None of them had a chemistry degree or formal experience. They got the recipe from books or off the Internet, and, if they made one error, their product would have resulted in numerous deaths. Ecstasy labs have been found in barns, mobile homes, motel rooms, houseboats, mini-storage units, apartments, and homes.

GHB is also known as the "rape drug" because it is odorless and colorless, and when dissolved in beverages, GHB causes drowsiness, loss of consciousness, and loss of inhibition. The precursor of GHB is gammabutyrolactone (GBL) , which is used as a degreasing solvent or floor stripper. Three men in Kansas City, Missouri, were recently indicted by a federal grand jury for participating in a conspiracy to illegally import GBL from Canada, which would have been used to make $1 million worth of GHB for distribution in the United States. The three men were charged in a 35-count indictment with charges ranging from conspiracy to import GBL to money laundering.

People using MDMA and GHB are often also using ketamine (Ket). Ketamine hydrochloride (HCl) is a cat tranquilizer that is also used in sexual assaults because it puts the victim in a frozen state for a period of time. It is also used legitimately in humans when prescribed by licensed medical personnel because it does not depress critical body functions in humans as much as other anesthetics. One Ket user when asked to describe its effects stated, "If you take enough (it will) give you a pre-view of your own death, put you in contact with seraph-like entities, and convince you that you have just seen God in a disco ball." Although most ketamine used in the United States is stolen from medical or veterinary practices, it is also manufactured illegally. One of the biggest raids on a Ket lab was in Tsuen Wn, China, where police seized $2.44 million dollars worth of the drug, which was being manufactured in block form. In addition to the ketamine, police seized a number of other drugs, a stun gun, 12 machetes, and three black hoods.

Because most of the illegal synthetic drugs sold in the United States are imported or stolen, there are a limited number of laboratories manufacturing these products. But there are a large number of crack cocaine and methamphetamine manufacturers. Cocaine is the world's most powerful stimulant of natural origin. Natives of South America have used the drug for at least 5,000 years, claiming that coca-chewing promoted clarity of mind and a positive mood. Traditionally, the leaves have been chewed for social, mystical, medicinal, and religious purposes. Unfortunately, the drug cartels have made cocaine in more potent forms available to drug users around the world. Most of the cocaine smuggled out of South America is cocaine HCl, an odorless, white crystalline powder with a bitter taste. It is usually snorted.

The use of freebase, or crack cocaine, is not new. The drug company Parke Davies introduced coca-cigars in 1886. Freebase is derived from cocaine HCl that has been chemically treated with ammonia or baking soda to free the potent base material from the salt. Crack/freebase cocaine cannot be easily dissolved in water, so it is usually smoked. The euphoric rush comes in just a few seconds, faster than an intravenous injection of cocaine HCl. Crack cocaine is highly addictive.

Crack addicts are known for their violence because use of the drug results in anxiety, fear, and paranoia. These are the people manufacturing the drug and running the crack houses. Occasionally the crackheads will find an old building with an inactive service, steal energy, and begin manufacturing their product. Because of their paranoia, they will not stay at the location for long before looking for another empty building with an inactive service. Crack manufacturers present two challenges for the revenue protection investigator and other utility employees. First, they may be stealing energy at the empty building to manufacture the crack cocaine. Second, if they are users, and they probably are, then there is a great potential for violence.

The illegal drug laboratories that present the greatest danger to utility employees are methamphetamine laboratories. Meth is a synthetic drug that stimulates the central nervous system. It is closely related chemically to amphetamine, but produces a greater effect. The drug's euphoric effects are similar to cocaine, but they are much longer lasting. Meth can be found in a variety of forms including pills, powder, and chunks and may be referred to on the street as crank, speed, and chalk. Pure methamphetamine hydrochloride, the smokable form of the drug, is called ice, crystal, glass, and quartz because of its clear chunky crystals. The drug can be smoked, snorted, swallowed, or injected.

Although there are several different ways to manufacture methamphetamine, most use the following ingredients (how many of these ingredients would you want to ingest into your body?):

- Ephedrine (cold and allergy medicine)
- Pseudo ephedrine (cold and allergy medicine)
- Alcohol (rubbing or gasoline additive)
- Toluene (brake cleaner)
- Ether (engine starter)
- Sulfuric acid (drain cleaner)
- Methanol (gasoline additive)
- Lithium (camera batteries)
- Trichloroethane (gun cleaner)
- Anhydrous ammonia (arm fertilizer)
- Sodium hydroxide (lye)
- Red phosphorous matches
- Iodine (veterinarian products)
- Sodium metal (can be made from lye)

- Methylsulfonylmethane (MSM; a vitamin supplement for animals)
- Table or rock salt
- Kerosene
- Gasoline
- Muriatic acid
- Campfire fuel
- Paint thinner
- Acetone

The only thing crazier that using meth is manufacturing it. Because of the chemicals used there is always a possibility that the lab will explode. In addition, the chemicals are highly toxic and just breathing the vapors can have terrible health effects. Short-term health effects may include eye irritation, lancination (shedding tears), redness/inflammation, and cornea injury. Inhalation may cause mucous membrane irritation in the nose and throat, and lung irritation resulting in cough, chest pain, and shortness of breath. There may be an accumulation of fluids and bleeding in the lungs, and a high concentration of vapor may cause skin irritation.

A utility employee could be exposed to the chemicals in a methamphetamine laboratory either through inhalation or skin exposure. The employee would be aware of the inhalation exposure because of the strong odor, which has been described as cat urine, ether, ammonia, acetone, and chemicals. An employee who may have inhaled fumes from a laboratory needs to get medical help immediately. The greater danger may be from skin exposure. Lab operators routinely dump the toxic waste left after manufacturing meth into streams, rivers, fields, sewage systems, or just in the backyard. A meter reader or other employee could come into contact with the waste and not know it until hours later. There is an additional danger to the investigator. If the investigator doesn't know the tampering discovered is at a meth lab and rings the front doorbell, there could be an explosion if there is a spark anywhere in the house when the doorbell rings.

The number of methamphetamine labs is increasing logarithmically each year, so it is imperative that all utility field employees know the warning signs that a lab is operational or may have been operational:

- Unusual strong odors
- Residences with windows blacked out
- Renters who pay the landlord in cash (Most drug dealers trade exclusively in cash.)
- Lots of traffic with people coming in and out at unusual times (There may be little traffic during the day but a heavy increase in traffic at night.)
- Excessive trash including large amounts of items such as antifreeze containers, lantern fuel cans, red-stained coffee filters, drain cleaner, and duct tape
- Unusual amounts of clear glass containers, including mason jars, being brought into the home
- Propane tanks

If you see any of these warning signs in the field, back off. Leave the scene immediately and, if you are a field employee, report it to your supervisor. If you are the revenue protection investigator, you may report it directly to the police or to your supervisor depending on your utility's policy.

## Indoor Marijuana Growing

The police and six local utilities in the York Region of Ontario, Canada, have been working together since 2001 to combat the proliferation of indoor marijuana grows in the area, recognizing that many of these operations are stealing energy. In 2002, police raided more than 1,200 indoor grows across the province, but they estimate there are approximately 12,000 grow houses in the region. The utilities recovered more than $2.6 million from these operations, an average of $2,167.00 per grow house. If the police are correct in their estimate of the total number that actually exist, then utilities in the province are losing more than $26 million each year to indoor marijuana grow houses. This means each honest rate payer in the area is paying an additional $34.00 per year to make up for the electricity the growers are stealing.

Known as *Operation York Connection*, the task force of police and utilities communicates that in addition to the financial burden this problem creates, there are other concerns and dangers:

- Overloading the power system can cause power outages for entire neighborhoods by blowing out transformers.

- Theft of electricity and tampering with electric wiring can easily lead to house fires.

- Haphazard and dangerous rewiring can lead to electrocutions.

- Restoring the grow houses back to a livable condition places upward pressure on homeowner insurance rates.

The communications to the public also mention some of the steps the utilities are taking to combat the problem and point out that the costs of these measures are also causing a drain on the following resources:

- Personnel to monitor power usage to determine fluctuations and potential illegal by-passes

- Line crews and trained personnel who monitor for potential electricity theft

- Trained personnel to perform the disconnects

- Personnel to repair cable and other damaged electrical equipment in grow house operations

- Administrative staff that support operations relating to each phase of the investigation process and the recovery of lost revenues

Some of the grow houses are small operations, run by the homeowner or renter. Others are connected to organized crime including biker gangs and ethnic criminal organizations. At one end of the spectrum there is the 69-year-old grandmother that pleaded guilty to turning her house in to a grow operation with 50 pot plants worth approximately $50,000. She was ordered to repay the utility $6,763 for the energy stolen.

In another case a 42-year-old man pleaded guilty to the cultivation of 3,500 marijuana plants but claims he was forced into the business by the Hell's Angels. He was ordered to repay $11,250 for the electricity stolen over one growing season. There is also evidence that Asian organized crime is responsible for the explosive jump in large-scale hydroponics marijuana grow operations across Canada and around the world.

The most important factor in indoor growing is light, and most growers use 400- or 1,000-watt high intensity discharge (HID) lighting specifically designed for this purpose. Growers must also control the environment including water, nutrients, and air. They use carbon dioxide generators and pumps for this purpose. In other words, they use a lot of energy. As stated earlier, a grow house will use at least 10 times the energy the building would normally use.

There is another potential problem at a grow house. They use a lot of energy and a lot of water, not a good combination if you bring them together. During a recent raid on a grow house the police were inspecting one of the grow areas when the revenue protection investigator arrived. The investigator pointed out to the officers that although the plants were on elevated platforms, there were a number of puddles under them with extension cords running through the puddles.

The utility should respond to the increase in marijuana grow operations with two goals or objectives. First, educate all field employees to recognize the signs of a potential marijuana grow house. Second, the revenue protection investigator should have a close working relationship with local law enforcement and be called to inspect the services at all grow operation investigations. You don't need to know about the raid in advance, but you should be called to the scene once it is secured by the police.

Signs that a house or other building may be a grow operation include the following:

- Strange people visit the house at strange hours.
- The residents are rarely seen.
- No one ever sees the moving van and furniture arrive.
- The blinds are drawn or the curtains are always shut.
- The lights are on but no one ever seems to be at home.
- Moisture or condensation forms on the windows.
- The roof is dry on chilly mornings.
- A humming noise, like a fan, is audible.
- The windows are covered with foil to preserve the heat in the home.
- The basement windows are boarded up.
- There is a lot of grow medium in the yard but no garden.

Another way to determine if you have a potential grow house is to test the harmonics on the electric line. The ballast used in the 400- and 1,000-watt HID lights has a specific harmonic signature. For detailed information on the harmonic signature, contact your local Drug Enforcement Agency (DEA) office or test your suspect site and several other sites and look for significant differences. This method of detection was tested successfully during a DEA case in the southeastern United States, but not all of their offices may be aware of the case.

A final note about drug labs and marijuana grow operations—they often have booby traps around them. Some of these are noisemakers meant to alert the operators when someone is approaching the building. Others can cause serious injuries or death. In Canada, police report that growers intentionally weaken the floor boards on porches so that the police will fall through them onto sticks with nails in them or other piercing devices. They have also used explosives, including hand grenades, chemical booby traps, and shotguns wired to pulleys so that they fire when someone opens the door.

# Defusing Anger and Aggression

Hopefully, the revenue protection investigator will not come across a drug lab or grow house every day, but the chances are that he or she will have to frequently interact with angry, and potentially aggressive, persons. Anger runs on a continuum of intensity that ranges from mild annoyance to rage. Some customers will express anger passively, whereas others will be more assertive. Almost every angry person wants the following:

- They want you to admit that you did something wrong.

- They want you to acknowledge and empathize with the pain and/or damage you have caused them.

- That want you to admit that you or the utility are 100% to blame for whatever happened and that they had little or nothing to do with it.

- They want you to take full responsibility by apologizing for what you did wrong, by offering to make amends, or by being appropriately punished for your misdeeds.

The most important thing to remember about anger is that it is a physical, not just an emotional, response. When a person becomes angry, certain chemicals are released into the bloodstream that increase the activation of the nervous system. The more chemicals that are released, the angrier the person will be. If you try to be rational with an angry person, or expect them to act rationally, you are doomed to failure. The best way to deal with an angry person is to get them to vent, to talk it out. To defuse the anger there are three general steps to follow:

1. *Acknowledgement of the emotion:* It is important to them that you know they are angry.

2. *Empathetic apology:* You don't have to accept responsibility, but you do need to let the person know that you are sorry they are in this position.

3. *Problem solving:* They want to know that this problem can be resolved or that it will not occur again, whichever is appropriate.

There are a number of factors that can cause and escalate anger, hostility, and aggression. Insecurity or a loss of control or predictability is a factor. You just caught this person stealing energy and they don't know what to expect as a result. Disrespectful behavior or attitude leads to anger. This includes any actions that are considered inappropriate in terms of the context or culture in which they occur. Lack of choices can result in anger. When someone's service is disconnected for nonpayment, they have one choice, to pay the bill. As we know, some of these customers select a second choice and turn the power back on themselves without paying the bill. Then the utility has to take away that choice.

Demonstrating respect to others, including utility thieves, is a primary means for avoiding or de-escalating anger, hostility, and aggression. If you answer anger with anger, the situation will escalate. Here are some factors to consider when dealing with angry people:

- Recognize that the angry person often feels threatened, anxious, or fearful.

- Focus on communicating respect with appropriate listening skills and nonaggressive body language.

- Attempt to establish some type of significance with the angry person's personal dignity.

- Remain calm yourself. Be a role model.

- Empathize with the person's position and the importance of his or her concerns.

- Watch what you say and how you say it.

- Sit down if possible.

- Let the person vent, responding with your active listening skills.

There are several things to avoid when speaking to an angry person. Do not blame the individual who is already angry. This will cause the person to become defensive and hostile rather than cooperative and understanding. Use the phrase, "Some unauthorized person tampered with your meter." Don't assume that everything you know about the case is correct. There may be other factors that you are not aware of. Listen to the angry person vent to determine if there are some additional facts you need to investigate. In one case, an elderly man was caught stealing and became enraged that the utility would accuse him of tampering with his meter. The investigator listened to the man and learned that his son was helping to pay his utility bill. Guess who the real thief was?

Occasionally you will not be able to defuse the anger and will have to disengage, meaning *get the heck out of there*. The goal of disengaging is to remove you from the threatening situation when it appears that all other efforts to defuse the anger and aggression have failed. Ideally, disengaging from an angry person involves an explanation of your behavior, allowing a cooling off period, and scheduling a time more conducive to resolving the problem or situation. You should disengage when:

- You have become angry yourself and are about to have a problem maintaining the active listening mode and the use of nonthreatening body language.

- You feel uncomfortable with the situation because of the rising level of emotion.

- You are concerned about your safety.

- You need time to further investigate the case or individual.

Defusing anger and disengaging require professional communication skills, including active listening. The revenue protection investigator should participate in communication workshops offered by the utility, utility associations, and local colleges.

## Summary

The revenue protection investigator should be prepared to deal with a dangerous account at every investigation. Always be alert as to what is happening around you and always have an escape route planned before you leave your vehicle. If you suspect that an account is a drug operation, back off. If the police call you to inspect the meters at a grow house, tell them you do not want to know the location. Give them a telephone number where they can call you once the area is secure.

If you doubt that dangerous accounts exist, go back to the beginning of this chapter and reread the examples. Those are actual cases.

## Recommended Actions

- Invite the local police department to talk to field employees about the drug laboratories and marijuana indoor grows in your service area. If you haven't already conducted or scheduled this briefing, pick up the telephone. This should be done annually.

- Consider forming a task force with the police similar to Operation York Connection. This concept has proved to be effective in identifying drug houses and revenues that have been stolen. It is also an effective means to get the public involved.

- Make copies of the warning indicators for a drug laboratory and the indicators for a grow operation, and provide these to all field employees and place them in all of the utility's vehicles. The lists should tell the employee who to call to report a suspect location.

- Coordinate with the police department so that the revenue protection investigator is called to all indoor marijuana grow investigations.

- Train all utility employees on how to defuse angry situations and how to disengage if necessary.

- Participate in communication skills workshops offered to the revenue protection investigator by the utility or other organizations.

# References

Ardell, Donald B. February 7, 2002. Defusing anger in others. *Seekwelllness*. http://www.seekwellness.com/wellness/reports/2002-02-07.htm (accessed July 23, 2004).

*Associated Press*. 2004. Fiji police raid huge meth lab (Article ID: Dq147338). June 9.

Cocaine Processing. http://cocaine.org/process.html (accessed July 23, 2004).

Dixon, Patrick. 1998. The truth about cocaine, crack, and heroin addiction. http://www.globalchange.com/drugs/TAD-Chapter%206.htm (accessed July 23, 2004).

Electricity Distributors Association (EDA) and local electricity distribution companies join with police force in "Operation York Connection." 2004. Electric Distributors Association news release, Toronto, June 4.

Graves, Todd P., and K. Harrisonville. 2004. Concordia men indicted for $1 million "date rape drug" conspiracy. Office of the United States Attorney Western District of Missouri news release, July 8.

KCI The Anti-Meth Site. Is there a meth lab cookin' in your neighborhood? http://www.kci.org/meth_info/neighborhood_lab.htm (accessed July 23, 2004).

Porrata, Trinka. Ketamine manufacturing centre smashed. Project GHB, 2002. http://www.info.gov.hk/gia/general/200403/24/0324245.htm (accessed July 23, 2004).

McWhirter, C. Cameron, and Jill Y. Miller. 2004. Meth stalks rural Georgia. *Atlanta Journal-Constitution*, June 6, pp. C1–C2.

Oregon Department of Homeland Security. Drug lab cleanup program. http://www.dhs.or.us/publichealth/druglab/meth.cfm (accessed July 23, 2004).

Porrata, Tinka. Project GHB. MDMA (Ecstasy, E, X, XTC, the Hug Drug). http://www.projectghb.org/ecstasy.htm (accessed July 23, 2004).

Project GHB. 2002. "What is GHB?" *Project GHB*, 2002. http://www.projectghb.org/what_is_ghb.htm (accessed July 23, 2004).

Sales Training International. Sales tip – defusing anger. http://www.saleshelp.com/guestservices/destinations/newsletter/DefusingAnger.htm (accessed July 23, 2004).

Schirch, Lisa L., and Dave Dyck. Module: Preventing and defusing anger and hostility. InnerAction/OFDA NGO Security Training. http://membres.lycos.fr/dloquercio/know-how/ressources.htm (accessed July 24, 2004).

Syndistar. 2001. What is ecstasy? http://www.inthezone.com/ecstasy/what.htm (accessed July 23, 2004).

www.streetdrugs.org. Methamphetamine labs. http://www.streetdrugs.org/methlabs.htm (accessed July 26, 2004).

# 8
# Investigative Challenges and Tools

## Internal Investigations

Who knows how to steal energy better than someone working for the utility, who not only has the knowledge but may also have access to meters and seals. In addition, this person probably knows the factors that initiate a revenue protection investigation and how to avoid attracting attention. Fortunately most employees are not involved in diversion or tampering, but a small percentage is stealing. Some of these employees are fixers, usually working on commercial meters.

An employee who is stealing utility services is not just another thief. This is someone who is stealing from his employer, and it is no different than if the employee were stealing tools or gas from the utility's gas pumps. Some employees steal just because they can. Others have poor attitudes at their jobs. These fall into three categories.

1. *Disillusioned* employees feel disappointed about things happening in the workplace. They complain but would probably not try to get even by stealing services.

2. *Disaffected* employees withdraw psychologically from others in the workplace. They are not motivated, complain a great deal, and may decide to steal services.

3. *Disgruntled* employees have strong negative feelings about the utility, management, and other workers. These employees are most likely to be stealing energy and to help others steal.

Supervisors are in the best position to identify employees in these categories. If a supervisor is having problems with a disaffected or disgruntled employee, this should be reported to the revenue protection investigator. However, before investigating the employee, the investigator should ascertain the relationship between the supervisor and employee to ensure that the supervisor does not have a grudge against the employee. The investigator should personally inspect the consumption records and the meter at the employee's house. This should not be assigned to a service person or other employee because of the sensitivity of internal investigations. To ensure that there is no conflict of interest, the investigator should not have supervisory responsibility over the employee being investigated. Obviously, they should not be related or best friends.

Larger utilities may have an internal ethics or employee hot line that result in tips regarding employee behavior. These tips are usually anonymous and require the investigator to assume that the accusation may be made by someone who has a personal conflict with the person accused.

There are other indicators that could trigger an internal investigation. Meters, seals, conductors, and other materials could be missing from the warehouse or from vehicles. A neighbor might report the employee. The employee may brag about stealing energy to another employee. If the employee is totally brazen, he might do what a lineman did at a large utility. He invited several line crews to his house for a cookout and ran service for the Japanese lanterns he had around the yard directly from his pad transformer.

The investigation of an employee for any potential wrongdoing, including theft of service, is a very sensitive and confidential undertaking. When preparing to initiate an internal investigation, consider the following:

- Ensure you have sufficient probable cause to initiate the investigation, and document it in detail.

- Review the bargaining unit agreement. Are there any provisions in the agreement or missing from the agreement that could affect your investigation?

- Keep the investigation confidential; very few people should be aware of it.

- Establish motive, if possible. Is this a disaffected or disgruntled employee, or someone who just wanted to get away with something?

- Determine if anyone else in the employee's work group should be investigated.

- Review the results of the investigation with legal counsel and Human Resources. (The final recommendations on the case will come from those offices.) The final decision will be made by management.

There are several things the investigator needs to know before initiating an internal investigation. Has the employee ever been informed that theft of services is a violation of the utility's policy? You may think this can be taken for granted, but not necessarily. The policy regarding theft of services by employees should be discussed in the new employee orientation and at an annual ethics briefing. Some utilities require employees to sign a statement at the end of the ethics briefing attesting that they understand the requirements discussed.

Collective bargaining agreements can affect the utility's right to investigate an employee. The National Labor Relations Board (NLRB) has ruled that investigative techniques the utility might use on union members who are employees are a mandatory subject of collective bargaining. In one case, an employer (not a utility) used hidden cameras for more than 15 years to detect employee theft, vandalism, and other wrongdoing. In pursuing the grievance of an employee, who was terminated based on evidence from one of these cameras, the union requested information regarding the hidden surveillance cameras. The union also requested that the company cease the use of the cameras without first negotiating with the bargaining unit. The company refused the union's requests saying they had not challenged their use for a number of years. The NLRB supported the union's right to the information and to bargaining, and the decision was upheld by the court. The NLRB recognized the company's right to protect the information on the hidden cameras, but also noted that these interests can be accommodated by bargaining with the union.

Do you have a collective bargaining unit at your utility? If so, review the agreement to determine if there are any provisions that affect internal investigations. The utility's counsel and the Human Resources Department will provide advice on any possible restrictions. There are cases such as one in which an investigator who was not aware of the bargaining agreement proved an employee had been stealing services. The employee was terminated. The employee filed a grievance based on the provisions of the agreement and was reinstated as an employee.

Until recently, under the Fair Credit Reporting Act (FCRA) if the utility used a third party to investigate an employee's behavior, it had to disclose the results of the investigation before taking adverse action. This did not apply to in-house investigations. The issue was addressed by the Fair and Accurate Credit Transaction Act of 2003, which excludes third-party investigations of employees' conduct from the requirements of disclosure and employee consent. Instead, employers are required to provide the employee with only a summary of the report after taking adverse action. The summary does not need to disclose the source. However, the limitation on disclosure of investigative reports is not absolute. The utility may be required to release the information during grievance proceedings. A terminated employee could bring legal action against the utility, and the information would be released during discovery.

A final note on the prevention and detection of employee theft of service. Some utilities inspect the meters of a certain percentage of employees each year. The meters are selected randomly, and the inspections are conducted by revenue protection investigators. Employees are informed of the random inspection program.

# Criminal Intelligence

A criminal investigation is undertaken when there is probable cause that a specific crime has been committed and facts are collected in relation to that crime. Criminal intelligence includes activities that help to uncover probable cause that will lead to an investigation. For example, assume your service area includes a large number of ice cream stores and there has been a major decrease in ice cream sales because of a new diet craze. You decide to review the consumption of energy used at all of the stores to determine if there are any trends that are unusual. The consumption has decreased somewhat at most of the stores, but you identify five locations where it has decreased considerably. Those five meters should all be inspected.

Law enforcement agencies collect intelligence using a number of different sources, including informants, undercover operations, and surveillance. To a lesser extent, the revenue protection investigator can use the same sources. Informants, for example include people who telephone the utility to complain that a neighbor turns the meter upside down on weekends. Undercover operations are not recommended for revenue protection investigators unless they are former police officers trained for these operations. They can be dangerous and there are procedures that law enforcement agencies use to protect undercover personnel that the revenue protection investigator would not have in place. A revenue protection investigator received information that someone was selling meters out of the back of a car at certain fast food restaurants. He had a description of the car and knew which restaurants the persons frequented. The investigator made a habit of driving by those restaurants until he spotted the suspect's car. He was direct. He told the suspect someone told him he could buy a meter from him and he wanted to know how much it would cost. When the suspect opened his trunk it was full of meters and seals. The investigator bought a meter and several seals. He also memorized the license plate on the car. The suspect was arrested later by the police. He still had a truck full of the utility's property.

In this situation, the investigator should have discussed the case with the police before going it alone. The suspect may also have been selling drugs and be a known violent offender. If so, he probably would have detected the deception, and the investigator could have been in trouble.

Don't conduct surveillance unless you are a former law enforcement officer with training and experience or a revenue protection investigator conducting the surveillance with a law enforcement agency. A number of things can go wrong when surveillance is conducted by unqualified personnel. However, a number of revenue protection cases have been solved using surveillance, usually when the utility investigator is working with a law enforcement surveillance unit.

There are a number of potential sources of criminal intelligence available to the revenue protection investigator:

- *Completed or ongoing investigations.* During an ongoing investigation, look for links from the suspect to others who may also be stealing—links between individuals and other individuals, between businesses and other businesses, and between individuals and businesses. Get below the tip of the iceberg. When you complete an investigation, go back through the information collected and look for these links. At the end of the revenue recovery interview, look the customer who you caught stealing in the eye and ask, "Who else's meter should I look at?" The answer may surprise you.

- *Newspapers and other media.* Follow the media to find out who was arrested for what. In one case, a string of gas stations was charged with tampering with the pumps to add the cost of an additional gallon to whatever was pumped provided it was not exactly 5 or 10 gallons. The people responsible for the scam knew that the state inspectors always measured 5 and 10 gallons. They made more than $1 million before they were caught. If you read about this case in the local newspaper you should inspect the meters at those gas stations. If the thieves were clever enough to program the gas pumps, they are probably clever enough to manipulate their meters. If they are doing one crime, they are probably doing other crimes. Also look for news on raids on indoor grow houses and other crimes that may also involve energy theft.

- *Informants.* A public awareness program will encourage informants to call. Most of these tips result in active investigations. A few will be from callers who are mad at neighbors and trying to cause trouble. People you have caught in the past may become informants if you treated them with dignity. If you think a past thief may have information on a current thief, call that person and ask for help.

- *Utility employees.* Employees can be a good source of information. They may hear things away from the office that need to be looked into. And, they could be the first to identify a fellow employee who may be stealing. Employees should be able to report possible irregularities to the revenue protection investigator knowing that their suspicions will be investigated and the information will remain confidential. If a customer is caught as a result of an employee tip, the investigator should provide personal feedback, again confidentially, to the employee. However, if an employee is caught, it might not be appropriate to provide personal feedback because that may violate the rights of the employee being investigated.

- *Law enforcement officers.* Cops are a great source of information. They already have informants and other sources of information, and some of the cases they are investigating, such as grow houses, may also be stealing energy—one more offense the criminal can be charged with. As you establish rapport with local police ask them to be on the lookout for information on energy theft. In one case an officer, who is a friend of the revenue protection investigator, stopped a car for a traffic violation. When the officer looked in the back seat he saw several utility meters. The driver was stealing meters and selling them at flea markets. This police officer was aware of the stolen meter problem as a result of a conversation with the utility investigator. Make sure the police in your area are equally aware. When you get a tip from an officer that results in an active investigation, give the officer personal feedback. It's the best way of saying thanks for the information.

- *Computer reports.* Your monthly exceptions report is an excellent intelligence tool. You can also do comparative analysis on target accounts like the ice cream stores. If you suspect that there are several people in a subdivision stealing services, do a comparative analysis of the energy consumption in the subdivision. Most of the lower users are probably using energy wisely. But a few could be the accounts you're looking for.

# The Scenario

The rest of this chapter discusses challenges and tools used to investigate more complex cases, which involve conspiracy to steal energy. The exploration of these challenges and tools is based on the following scenario.

While reading the meter at 1520 Salsbury Street, meter reader James Burton noticed that the disk was not turning but that a heat pump next to the meter was operating. On closer inspection he found that a hole had been drilled from the inside of the house, through the meter base, and into the meter. A piece of wire was inserted into the hole to stop the disk from turning. The external plastic seal was in place and did not appear to have been tampered with. Burton radioed this information to the dispatcher and continued reading his route.

The information was forwarded to revenue protection investigator Mary Howell. Mary checked the Customer Information Services/Customer Relationship Management (CIS/CRM) file on the account and found that it was in the name of Charles and Joyce Jones. The service was transferred into their names three years ago. There was a 15% drop in consumption beginning eight months ago. While searching the CIS/CRM, Mary learned that Charles Jones is also the customer of record for the Clean Sheets Motel. When comparing the electric consumption at this motel with similar motels, it appears the Clean Sheets is either very energy conscious or that there is a potential problem at this location.

Mary conducted the crime scene investigation at 1520 Salsbury Street. The scene was as reported by the meter reader. Mary removed the seal and pulled the meter. What appeared to be a straightened wire coat hanger was sticking through the hole in the meter base and into the meter. Mary collected as evidence the seal, the meter, and the wire. She placed a cover on the meter socket with an investigative sticker. No one answered the door so she left a door hanger stating that there was a safety problem at the meter and the Joneses should call her office when they returned home to discuss the situation. She documented the crime scene investigation using her digital camera.

Mary then met Walter Fry, meter technician at the Clean Sheets Motel, located at 14 Main Street. It appeared that the test switches at the service had been tampered with and that not all of the energy used was being metered. Walter corrected the situation, but because there were no obvious safety problems or tampering, the service was not disconnected.

Upon returning to the office, Mary worked with the billing department to determine that the estimated loss at the Jones' home was $550.00. The estimated loss at the motel might be as high as $8,000.00. She decided to monitor the meters and the usage at the motel for three months before determining if the location had been underbilled.

Three days passed and Mary had not received a telephone call from the Joneses. (Perhaps they are on vacation?) She had a service representative check the location, and he reported that there were jumpers in the meter socket. Apparently the seal had been cut, the cover removed, metal bar jumpers placed into the socket, and the cover replaced. The seal was bent to make it look like it was intact.

Returning to the scene, Mary collected the seal, cover, and jumpers as evidence. She placed a new cover in place with a locking band and security lock. Again, no one was home so she left a door hanger with her telephone number. Two hours later she received a telephone call from Mr. Charles Jones. He said he just returned home and found her door hanger. He wanted to know what the problem was and when his lights would be turned on. Mary suggested that he come to the office to discuss the situation. He said he would be there in one hour.

He arrived 90 minutes later with Mrs. Jones and his brother Paul. Paul claimed he is a partner in the motel and had to drive Charles and Joyce to the utility because they were too upset to drive themselves. Mary asked Paul to wait in the lobby while she discussed the situation with the Joneses. He protested but then agreed and left the room. Mary then asked to see the Joneses driver's licenses to make sure they were Charles and Joyce Jones. Mary told them what was found and presented a bill. Pointing out to them that they are prominent business people in the community, she ensured them that the utility wants to clear up the matter. They agree, although they claimed they did not know who tampered with the meter. Charles wrote a check for the total amount including meter tampering and cut seal charges.

Within three days following the Jones interview, meter readers and service representatives found five more cases where a hole had been drilled from inside a building and into the rear of the meter base and meter. In all of these cases, the tampering appears to have begun six to eight months ago. One of these meters was at the home of Paul Jones. In all of the other cases, the customers claimed they didn't know anything about the incident, but they were eager to pay the back bill and additional charges and to have their service restored. Paul Jones did not contact Mary but an inspection of his service three days after it was disconnected found the meter base cover, locking band, and lock all in place. It appeared that he had temporarily moved to the Clean Sheets Motel.

Other information learned since the Jones' interview includes the following:

- The Jones have a third brother, John Jones, who lived on the same system. John lives at 1422 Prince Street and owns the Great Food Deli at 203 Willis Street.

- Joyce Jones has a sister, Lucy Ringo, who is married to George Ringo. George is a contractor and licensed electrician who works out of his home. The name of his company is Ringo Electric. His meter did not appear to have been tampered with, but his consumption was low. A check meter has been placed on the pole at his service and will be monitored for a month.

- Charles Jones owns several rental properties with William Smith under the name of Jones and Smith Properties. Two of these properties had meters tampered with using the "drilling from behind" method.

- George Ringo has two brothers on the same utility system. Frank Ringo is constantly in trouble with the police for minor offenses. Frank is employed as a mechanic at a local car dealership. Simon Ringo served time in jail for selling narcotics, but was released last year. He sometimes works with George but has no other visible means of income.

## The Challenge and the Tools

Mary collected all of the charges from the Joneses for the tampering at their house, but she was still investigating the motel. In addition, it seemed that a fixer had been working the area using the same method to steal services at all of the residents where this method was detected. How many other residences had the fixer visited, and had he also tampered with meters at businesses?

What started as a single theft at a residence grew to a conspiracy investigation thanks to Mary's investigative prowess. At this point the investigator could be overwhelmed by the amount of information available, the need to decide what to do with the information, and the opportunity to develop additional information.

The postinterview information, which is numbered, is referred to as "Information Inputs." It is developed from a number of sources including utility records, a courthouse search for home ownership and business records, information from law enforcement, and other available sources. Information inputs should be listed and numbered. Two questions should be asked at this point in the investigation. Where are we in the investigation? Where should we go from here? Because this has become a conspiracy investigation involving a fixer, the police or an investigator from the prosecutor's office should be told of the investigation. To help answer these questions and to help illustrate the case to the police or prosecutor, the investigator might use all or some of the following graphic investigative tools:

- Investigative flowchart
- Social network analysis
- Investigative matrix
- Link analysis

Investigative flowcharts show the pattern of activity in the case in the sequence in which they occurred. They help to illustrate the flow of events, and, in some investigations—such as drug investigations or money laundering—they are used to show the flow of a commodity. Investigative flowcharts are used in case planning, case evaluation, and courtroom presentations. Flowcharts may also be used as part of the documentation for a utility commission complaint.

When preparing an investigative flowchart the following conventions should be followed:

- Events are represented by circles with the date of the event next to the circle.
- People are represented by boxes with the name(s) inside the boxes.

- Events and people are connected by a solid line in the order of the events.
- The sequence of events is illustrated from left to right.

Figure 8–1 illustrates an investigative flowchart for our scenario.

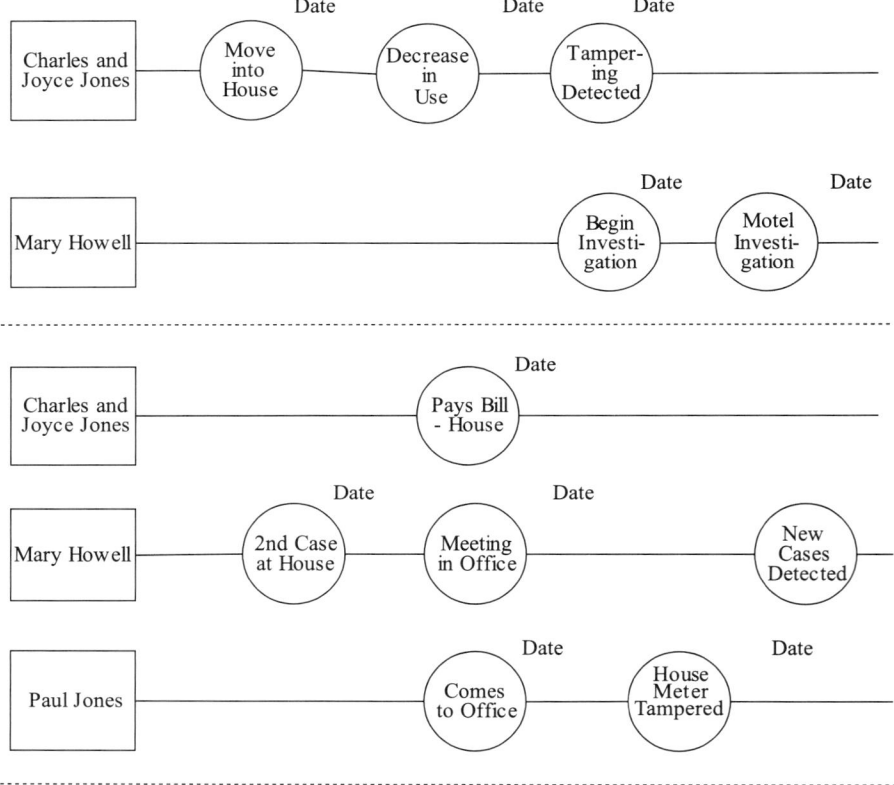

Fig. 8–1. Investigative flowchart

The investigative flowchart in figure 8–1 illustrates the beginning of the investigation. As the investigation of the individuals identified in the information inputs continues, the flowchart will expand. When the investigation is completed, the flowchart will illustrate the complete case.

Social network analysis is an investigative tool that illustrates direct relationships between people. It is especially valuable in complex fraud investigations as well as theft of service conspiracies. A social network analysis also has standard conventions, which are different from the flowchart conventions:

- Individuals are represented by names in circles.
- Known associations or relationships are represented by solid lines.
- Suspect associations or relationships are represented by broken lines.

Figure 8–2 illustrates the social network analysis for our scenario and includes the persons identified in the information inputs.

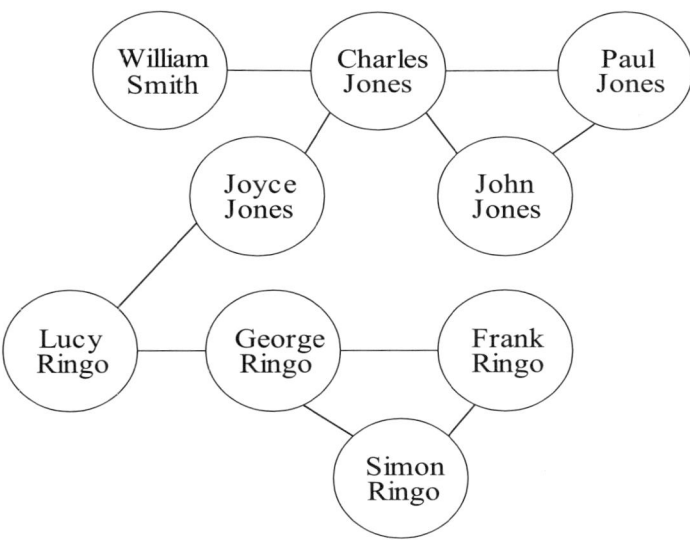

Fig. 8–2. Social network analysis

In our scenario, a social network analysis would be used to identify other meters that should be inspected, including William Smith, John Jones, and each of the Ringo brothers. When used in fraud investigations, this analysis helps to identify individuals who are participating in the revolving door game. It may also identify someone playing the name game if several of the connections lead to the same circle where different names are used.

An investigative matrix helps identify relationships between people, between businesses, and between businesses and people. It is also used to develop the link analysis diagram. As with the other tools, there are conventions for creating and illustrating the investigative matrix:

- Collect as much information as possible and put it into the information input format.
- Read the information and highlight the names of the individuals.
- Read the information again and highlight the names of businesses and other organizations in a different color.
- Alphabetize the names of the people.
- Alphabetize the names of the businesses and organizations.
- Develop the matrix by listing the people alphabetically. Then list the businesses alphabetically.
- Where there are known associations, place a solid circle in the box connecting the two entities.
- Where there are suspected associations, place an open circle in the box connecting the two entities.

Figure 8–3 illustrates the investigative matrix for the scenario.

In the scenario all of the associations are known, so we have solid circles. The possible suspect association is the link between Simon Ringo and Ringo Electric. Our information input says that Simon may work part-time for his brother's business, but that hasn't been confirmed.

The information in the investigative matrix is used to prepare the link diagram. Link diagrams have their own conventions:

- People are represented by circles.
- Businesses and organizations are represented by squares or rectangles.
- Known associations are linked with solid lines.
- Suspect association are linked by broken lines.
- Circles representing people who are part of an organization are placed inside the organization rectangle.

**Investigative Challenges and Tools**

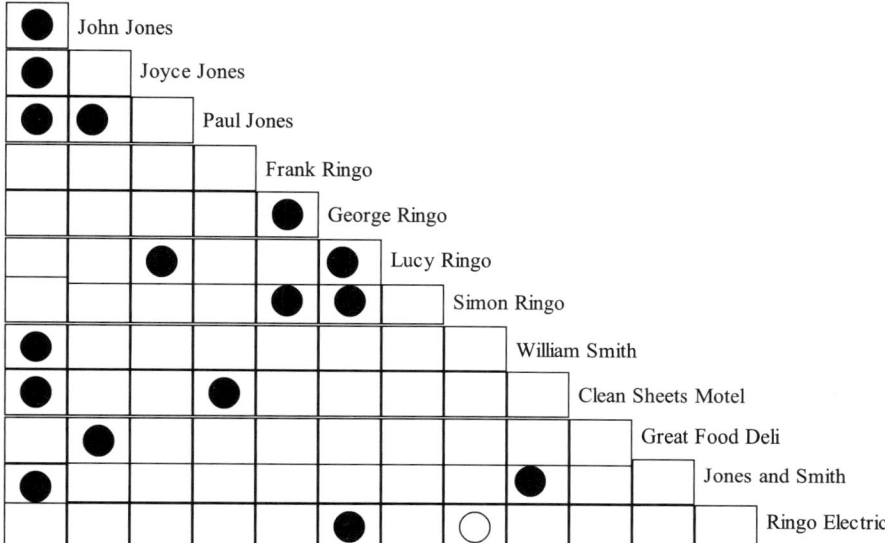

Fig. 8–3. Investigative matrix

- If an organization is the sole enterprise of an individual, its rectangle may be placed inside of the person's circle.
- If a person is part of more than one organization, that person's circle is placed in the intersection of both organizations.
- Preferably, lines on the link diagram do not cross.

Let's look at some examples of the conventions used in link diagrams (fig. 8–4).

Figure 8–5 illustrates a link analysis diagram for the scenario.

The link analysis diagram helps to illustrate the connections in this case. We know that we found tampering at Charles Jones' and John Jones' houses. There is possible tampering at the motel. Jones and Smith own several properties where tampering has occurred. So the next steps in the investigation are to

- Try to identify the other houses owned by Jones and Smith Properties and inspect those meters.
- Inspect the meters at the Jones and Smith office and the Great Food Deli.

*More than one person in an organization.*

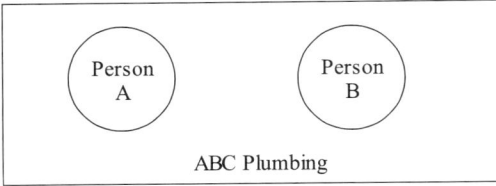

*Person linked to more than one organization.*

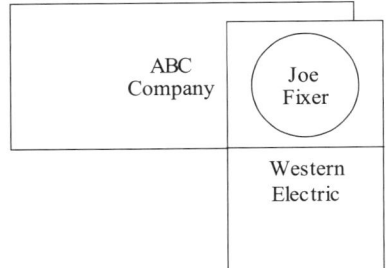

**Fig. 8-4. Conventions for Link Diagramming**

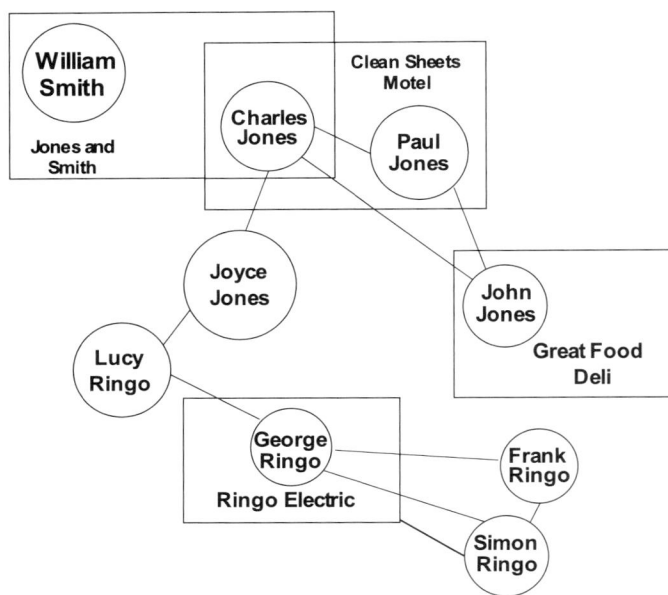

**Fig. 8–5. Link analysis diagram**

- Inspect the meters at the houses of John Jones, George Ringo, Frank Ringo, and Simon Ringo.

- Find out if any of these people own additional properties or have an interest in other businesses and inspect those meters.

- George Ringo is an electrician and our prime suspect as the fixer. Simon Ringo may assist him on occasion. Try to find out who are the clients of Ringo Electric and check those meters.

- Check with the police to find out if they have information on any of the people involved in the investigation. Is there word on the street that George Ringo can help you save money on the electric bill?

- You may also want to check the meters at the car dealership where Frank Ringo works.

Investigative flowcharts, social network analyses, investigative matrixes, and link analysis diagrams are not tools the revenue protection investigator will use on a daily basis, but they should be in the investigator's toolbox. These are great tools to use when you believe a conspiracy investigation is developing. They help to define the direction of the investigation, and they are effective tools at prosecution. At trial, the investigative flowchart illustrates to the court the sequence of events, and the link analysis diagram illustrates the relationships between all of the entities identified during the investigation.

In the mid-1980s a seasoned revenue protection investigator attended a certification workshop and learned to use these tools, thinking he would probably never use them. Two years later he found himself involved in a conspiracy investigation where a group of fixers were targeting commercial and industrial accounts. He used the investigative tools discussed in this chapter to present and illustrate the case to the state police and to the prosecutor. As a result, the fixers were convicted and a significant amount of revenue was back billed.

There are a number of software programs available to assist you in using these tools. The two programs used to develop the initial figures in this chapter were SmartDraw 6 and RFFlow 5.

## Summary

Before you begin an internal investigation, review the suggestions at the beginning of this chapter. Catching an employee stealing energy and then having that employee reinstated is not only embarrassing, but it also has a negative impact on other employees' perceptions of the revenue protection program.

Develop an ongoing criminal intelligence capability. This can be as simple as paying more attention to the local news, using the information in the computer system to do comparative analyses, or developing informants. Use your liaison with law enforcement as an intelligence source.

Be alert for potential conspiracy investigations. If you have fixers working in your service area, you have conspiracies. Use the investigative tools presented in this chapter to analyze the information collected during those investigations and then discuss the case with police or an investigator at the prosecutor's office.

## Recommended Actions

- Supervisors at your utility should report concerns over disgruntled employees who may be stealing energy.

- If you have a bargaining agreement, discuss the agreement with counsel and/or human resources to identify possible restrictions on internal investigations.

- Some type of criminal intelligence capability should be in place. This capability makes the revenue protection program more proactive. You are not just *responding to* tampering reported in the field. You are actively looking for possible cases.

- If you have a case that involves several parties, collect all the information you can on those parties and prepare a list of information inputs. Do you have enough information to use the investigative tools presented in this chapter?

# References

Goemaat, J. 2004. A case of good judgment: Here's what companies need to know before conducting investigations into employee conduct. *Security Management.* http://www.highbeam.com/library/doc3.asp?ctr1Info=Round5%3AProd%3ADOC%3APr (accessed July 27, 2004).

# 9
# Working with Law Enforcement and Going to Court

## Establishing Liaisons

Forming liaisons with law enforcement officers, fire investigators, and codes inspectors will greatly enhance your revenue protection program. Liaisons with public housing and drug abatement response teams are also valuable. The time to form liaisons with law enforcement and other agencies is before you work the first case with them. They won't even know that the utility has a revenue protection program until you contact them and explain the potential economic impact and physical dangers of current diversion and tampering. Also explain that some of the people committing utility theft are committing other crimes as well.

Your utility should have specific decision factors that dictate when to work with law enforcement on current diversion investigations. Some utilities call law enforcement every time. The police conduct the investigation and collect the evidence. They then refer the case for prosecution.

Let's begin by establishing liaisons with local law enforcement. How much effort this requires will depend on the size and number of law enforcement agencies in your service area. If you are working with small departments, start at the top. Meet with the police chief or sheriff and provide information on the revenue protection problem and your utility's approach to control losses. If you are working with larger departments, you will meet with the person in charge of operations and possibly the chief investigator. Your goal is not just to meet with key people but to form working relationships with the officers and investigators who are on the street. A good way to reach these people is through the department's training officer.

Almost every law enforcement agency has someone who is in charge of developing an annual training schedule for the department. At least one day will be spent on the firing range and another updating laws and procedures. If you can get on the training schedule to present a module on current diversion, you will create an awareness of a crime about which many of officers may be unaware. More important, you will get to meet all the department's operations personnel. They will know what to look for at crime scenes and who to call if they find something suspicious at the meter.

Make sure you meet with your local drug task force. This relationship should be so close that they call you to the crime scene when investigating indoor growing operations. If you are asked to accompany them to the crime scene, remember two important rules:

1. You do not want to know in advance the location where the warrant will be served. Occasionally the drug investigators arrive to find the building empty because someone tipped off the grower. If you don't know the location, you won't be suspected of telling the bad guys investigators were coming. Either meet the drug investigators at a neutral point en route to serving the warrant or be on call so they can radio the location to you once they have arrived and cleared the area.

2. That's the second rule. Never approach the building until the police have complete control and all suspects in the area have been apprehended. Guns are common at all drug operations, and you don't want to be in the middle of a gun battle. Also, if they are serving the warrant at a meth lab, these have a tendency to catch on fire or explode. Stay at least one-half mile from the scene until it is cleared by police.

You should also liaison with the fire department. Some fires first believed to have been accidental electrical fires are actually the result of meter tampering or current diversion. Some have resulted in serious injury or death in addition to significant property loss.

A consultant was contracted to evaluate fires resulting from product failure or misuse, focusing on fires that started from the receptacle back to the service. The study did not include extension cords, Christmas tree lights, or the like. Because the study is conducted every several years, he asked to see the case files from previous studies. After reviewing a number of case files, the consultant determined that as many as 30% of the fires whose point of origin was determined to be at or near the meter could have been the result of energy theft. But the cause was determined to be product failure or misuse because the fire investigators were never trained to look for tampering or diversion.

Remind the fire departments in your area about the fire that started in an old house when the residents ran an extension cord from the house next door after being disconnected for nonpayment. Two firefighters died trying to suppress the fire. Tell them also about the girl who died in a house fire caused by the straps her father had put behind the meter. Fire departments play an important role in combating power theft.

The fire department will also have a training officer, but firefighter training is often done using videos or CD-ROM/DVDs. This is because many departments are volunteer fire departments where the firefighters meet once or twice a month for training after their regular jobs. Even in full-time departments, training has to be scheduled around other activities, including fighting fires. Also, the training culture is different. Many of the training programs used by fire departments are produced by the National Fire Protection Association (NFPA). NFPA produces excellent training, usually in a video or CD-ROM/DVD format.

If you are working with a large fire department, try to schedule meetings at the major fire halls. Keep the presentation brief. If you are working with volunteer departments, find out when their training sessions are scheduled and try to get on the agenda.

If your utility has the resources, consider producing a videotape that can be used for training by both police and fire departments. The training video usually consists of three parts:

1. An office video of a utility investigator at a desk or table. Explain the seriousness of the problem and the potential consequences of energy theft. Provide several examples and then describe the roles

of the utility investigator, police officer, and firefighter in combating the problem.

2. A field video showing a correct residential meter center and then showing what to look for if tampering is suspected. Don't go into too many methods used to steal energy. Concentrate on checking the seal, inverted meters, and other obvious indicators. Then go through the same approach with commercial metering.

3. The final part of the video can be in the office or field and emphasizes the need to work together to control theft of service, which can result in increased utility rates and public exposure to dangerous situations. End by giving them your contact information.

Be sure to liaison with codes inspectors as well as electric and building inspectors. These people also need to know what to look for, and they can be valuable allies in many situations. If possible, also get to know the judges and the clerk of courts in your service area.

You need to provide feedback to the liaison system once your relationship is established. Some utilities share their monthly revenue protection reports with these public safety agencies. The reports list individual cases by account number—no names or addresses—the disposition of the case and the amount of revenue recovered. If you have worked a major case with a law enforcement agency, the people you worked with deserve personal feedback. Keep them posted on the progress of the case.

Another feedback approach is an annual meeting hosted by the head revenue protection agent. In smaller utilities the general manager may also be present. Invite personnel from all the agencies and give them a short update on the success of your program. Ask for suggestions on how the program can be improved and thank them for their assistance. Remember to keep the meeting short; they are as busy as you are. Scheduling the meeting as a 7 A.M. breakfast often works, provided the meeting is over in approximately 45 minutes. Invite the following people to the breakfast:

- Representatives from local law enforcement agencies
- Representatives from local fire departments
- Prosecutors

- Judges
- Clerk of courts
- Codes inspectors
- Investigators from other utilities that provide service in the area

Work with the other utilities in your area. One midwestern electric utility in the United States recently developed an innovative program with the cable television franchise. Each company is cross-training the field personnel from the other company on how to detect electric or cable theft. Both companies have significantly increased the number of eyes and ears they have in the field.

## Solvability Factors

Like most organizations, law enforcement agencies don't have all the assets they would like to have, including personnel. And in today's world where many of them are dealing with the overwhelming drug problem and the threat of terrorism, utility theft is not going to be a high priority. You can increase the priority of your cases somewhat by understanding the criminal justice system numbers game, which is based on the fact that approximately 95% of all revenue protection investigations are solved, 95% of those that go to court result in convictions, and seldom if ever is the conviction overturned if appealed. That means that a police officer investigating a revenue protection case will almost always solve it. The prosecutor will almost always win the case if it goes to court. And the judge has an extremely low probability that the case will be appealed or overturned. Everyone looks good!

Why are most revenue protection cases solved and successfully prosecuted as compared to other types of crimes? The reason is related to *solvability factors*. Because they have limited investigative resources, law enforcement agencies commit those resources to horrendous crimes. They also commit lesser resources to crimes that have a high probability of being solved. Criminal justice research has examined the differences in the elements present in crimes that have a high probability of being solved compared with those with a low probability. The more factors that are present, the higher the probability that the crime will be solved. Let's apply these factors to the typical revenue protection investigation.

- *Witness to the crime.* In revenue protection cases, there is usually the meter reader who discovered the problem and the investigator who arrived at the scene. There may also be a law enforcement officer or neighbor, although neighbors rarely want to testify if the case goes to court.

- *Knowledge of the suspect's name.* You either have the name on the account or you can do an investigation to find out who lives at the account. Drive by and get the vehicle license plate numbers at the location or check with other utilities and service organizations to ask what name they have for the account at that location.

- *Know where the suspect can be located.* This is usually the location where the tampering or diversion occurs. If not, you can conduct a paper chase to find out who owns the property.

- *Description of the suspect.* A description of the suspect is preferred but may not always be needed. You want the name so that you know who benefited from the theft and can attempt to recover the revenue.

- *Identification of the suspect.* This is a combination of *know where the suspect can be located* and *description of the suspect*. This also means having a Social Security Number (national identification number) or other specific identification information.

- *Lost property that is traceable.* This is often a guess unless the customer was an inactive account who removed the boots from your meter and registered the energy used. An analysis of the account can usually provide a fairly accurate estimate of the energy not metered as a result of the diversion.

- *Significant method of operation (modus operandi or m.o.).* What did the customer do to divert energy? You should have photographs or digital images to show the specific method used.

- *Significant physical evidence.* You have the meter that was tampered with, the seal that was cut, and other physical evidence depending on the method used.

- *Description of the suspect's vehicle.* This is usually not needed in a current diversion case unless it becomes a police investigation.

- *Positive results from the crime scene search.* This is your physical evidence, eyewitness testimony, and photographs or digital images. For this evidence to be presented in court, it must be collected and preserved according to the rules of evidence.

- *Crime can be solved with additional effort.* Additional effort is not usually needed unless you are investigating a fixer or other major revenue protection case.

- *Suspect has motive and opportunity.* The suspect's motive is obvious. Opportunity is also obvious.

When the majority of these solvability factors are present, the crime, in this case current diversion, has a 95% probability of being solved and successfully prosecuted.

Other utilities call an officer to the scene at all investigations, but the utility conducts the investigation and collects the evidence. The officer is there to provide protection for the utility investigator and he or she will serve as a witness if the case goes to court. Most utilities call the police only for special investigations. These include possible drug-related operations, potentially violent customers, or investigations in high-crime areas.

# Major Crime Scene Investigations

Let's assume that the police have obtained a search warrant for a major indoor marijuana grow where the criminals are thought to be stealing energy. You have been invited to be a part of the major crime scene search and although the police department will be in charge, it is important that you understand your role on the investigative team. These are the usual steps to prepare for when conducting a major crime scene search:

- Establish the basic premise for the search.
- Prepare for the search.
- Approach the search area.
- Secure and protect the scene.
- Conduct a preliminary survey.
- Evaluate physical evidence possibilities.
- Write a narrative of exactly what takes place during the search.
- Photograph the scene and all the evidence.
- Prepare crime scene sketches as appropriate.
- Collect physical evidence.
- Conduct the final survey.
- Release the survey team.

The search strategy and the best time to serve the warrant will be determined as you discuss *basic premise* of the case at hand. During the *preparation* stage, the command structure and lines of communication will be established. Your specific role as the utility investigator will be defined by the law enforcement officer in charge of the search.

As you *approach* the crime scene, be alert for evidence relating to possible current diversion and take copious notes. As the utility expert, you should look for unsafe conditions that could endanger the search team. In one case the grower had electric wires lying on the floor below the plants in puddles of water!

The police team will *secure and protect* the area. You should not go the to crime scene until after the warrant has been served, the scene is secured, and all suspects are under police control. Uniformed police officers are usually present to keep unauthorized personnel from getting into the crime scene.

The *preliminary survey* consists of a walk through the entire crime scene. Someone will begin to record the search narrative and the police photographer will take the initial photographs. Begin to take notes on the current diversion investigation and photograph evidence relating to your part of the investigation, but do not touch anything or collect evidence until the police officer in charge gives permission. The police photographer will want to record the same evidence that you have photographed.

Working with your police counterpart, *evaluate physical evidence possibilities*. You are the expert on utility theft and will explain to the officer what you find and its effect on the service metering. The physical evidence possibilities depend on the methods used to steal the energy.

Develop a written *narrative* of your part of the investigation. Be sure to record the time, date, and location as well as all facts pertaining to the tampering or diversion. Also record the names of the police officers you are working with.

Take *photographs* of the metering for your records even though the police photographer will also take photos. Crime scene *sketches* will be completed by the police specialist and are not usually prepared by the utility investigator. Sketches are used to supplement the photographs.

Working with the police officer assigned to you, begin the *physical evidence collection*. The police will tag and bag the evidence and establish and maintain its custody. Your role is to describe each piece of evidence and its relationship to the utility theft investigation.

During the *final survey* you will make sure that you have conducted a complete investigation, collected all the possible evidence, completed your notes for the final narrative, and recorded the final scene exactly as you leave it; include your final photographs. When the officer in charge is satisfied that the entire crime scene search is completed, he or she will *release* you and the rest of the search team from the scene. There will be a follow-up meeting to discuss the search, the evidence collected and to decide where the case will go from this stage. Major crime scene investigations take time. You may spend the entire day at the scene.

# Preparing for and Going to Trial

In most cases the revenue protection investigator will be responsible for preparing the case file. The file should include

- One-page executive summary recounting the history of the case
- Investigative reports
- The customer's application for service and other account documentation
- Billing and payment records
- Documents collected during the paper chase
- Transcripts or summaries of interviews
- Photographic log and photos
- Evidence log

As an investigator you may be called to provide testimony at pretrial hearings, civil cases, and criminal prosecutions. You will also be interviewed at depositions that are held prior to hearings or trials.

A *witness* is defined as, "One who gives evidence in a cause before a court and who attests or swears to facts or gives or bears testimony under oath" (*Handbook of Forensic Services*). A witness either presents *eyewitness testimony* or *opinion evidence*. If presenting eyewitness testimony, the witness can only attest to what the senses took in. What did you see, hear, smell, or feel? When presenting opinion evidence, the witness can make assumptions and reach conclusions based on the other evidence presented.

To be able to present opinion evidence you must first be qualified as an *expert witness*. When you take the stand the attorney for your side will ask for your qualifications regarding the matters before the court. The attorney should have your résumé so that your work experience and education can be entered into the record. The attorney will then ask the judge to qualify you as an expert in your particular area. The attorney for the other side has the right to challenge your qualifications, but this seldom happens. Once you are qualified as an expert, the attorney will begin to ask questions relating to the case. As an expert witness you can now develop and present conclusions based on the other evidence presented. Remember that you have been qualified as an expert in a very narrow area of expertise. You're not supposed to be an expert in all areas.

A deposition is a formal pretrial statement at which you present the same evidence you will present at trial. Attorneys for both sides and a court reporter will be present. The basis for the preparation of a deposition is good report writing because you will refer to your report throughout the deposition. The report you take to the deposition should include

- Suspect information including a physical description if you have one
- Notes on any statements made by the suspect
- Names and statements of any other witnesses
- A list of physical evidence including where it is located and its relevance to the case
- The names of investigators and others who were at the crime scene

If you are referring to notes during the deposition, and you should, bring two extra copies. One copy is for your attorney and the other copy is for the customer's attorney.

Keep in mind that a deposition is a formal pretrial statement. Maintain proper demeanor at all times. Firmly request a copy of the deposition so that you can review it before you go to court. If there are any errors, omissions, or other irregularities in the deposition, contact your attorney and explain them. Take a copy of your deposition to court, and don't hesitate to refer to it before answering questions while on the stand. Copies will be on the tables of both attorneys.

There are some general areas that you should be aware of before being a witness at trial or at a hearing:

- *Appearance.* Appearance is always important. It provides the judge and jury (if there is a jury) with the first impression of the investigator as a witness. Don't wear jeans with holes in them, but don't overdress either. When a utility employee arrives in court in expensive clothes and shoes, people in the courtroom begin to wonder if their utility bill is too high. In general, dress as you would for work.

- *Invoking the rule.* The judge may instruct the witnesses that they are not to discuss the case with anyone other than the attorneys for each side. You may also have to leave the courtroom so that you don't hear the testimony of other witnesses.

- *Taking the stand.* Act professionally the entire time you are in the courtroom. When called to testify, the investigator should go to the clerk to be sworn in and then take the stand.

- *Direct examination.* Direct examination is the questioning by the prosecutor. This may have been discussed in a pretrial conference. If the investigator is going to get down from the stand to show the jury a piece of evidence or to demonstrate some act such as installing a meter, that act should be practiced before going to court.

- *Cross-examination.* This is the most uncomfortable time for the investigator. The defense attorney will try to tear apart the carefully imparted facts presented during the direct examination. The defense attorney may also try to discredit the investigator. The investigator should continue to act professionally and remain fair and impartial.

- *Objections.* The prosecutor and the defense attorney will both make objections during your testimony. When they do so, do not answer the question until the judge rules, and then only answer if instructed to do so by the judge. If the judge rules that the objection is *sustained*, the objection is valid and the witness may not answer the question. If the judge rules that the objection is *overruled*, the witness will answer the question.

- *Jury deliberation.* If there is a jury trial (rather than the judge adjudication) the judge will instruct the jury to weigh the evidence. Some of the things the jury will consider are:

    ○ Did the witness seem to have the opportunity to see and to know the things about which the witness testified?

- Did the witness seem to have an accurate memory?
- Was the witness honest and straightforward in answering the attorney's questions?
- Did the witness have some interest in how the case should be decided?
- Did the witness's testimony agree with the other testimony and evidence in the case?
- Has the witness been offered or received any money, preferred treatment, or other benefit to get the witness to testify?
- Has any pressure or threat been used against the witness to affect the truth of the witness's testimony?
- Did the witness at some other time make a statement that is inconsistent with the testimony given in court?
- Was it proved that the witness had been convicted of a crime?
- Was it proved that the general reputation of the witness for telling the truth and being honest was bad?

You may present evidence and exhibits while on the stand. If you took the crime scene photographs, these will be placed into evidence while you are on the stand. The same procedure will be followed with other evidence including the meter and seal. Several exhibits may be introduced in addition to the evidence. These include the back bill and the method used to compute it, as well as the meter test report.

The utility may present creative exhibits. For example, if the customer was stealing energy by bypassing the potential circuit, you could present an exhibit that includes a meter base with a light on the load side. You could then demonstrate that when the potential clip is in the closed position, the light is on and the disk is turning. When the potential clip is bypassed, however, the customer continues to receive energy, but the disk is not turning. The energy used is not registered on the meter.

Here is a good witness checklist to follow to ensure that your testimony is professional and believable:

- If you have never been to court, visit the courtroom for several hours before you are called to testify. See the court in action.
- If possible have a pretrial conference with the prosecutor. If the prosecutor is unavailable, discuss the case with the utility's attorney.
- Have good notes and/or a report and bring four copies to court. You need a copy for reference; and as soon as you refer to your notes, the attorneys for both sides will want, and they have a right to, copies of the notes. The fourth copy is for the judge.
- Be at court on time. While waiting for your case to be called, be alert and maintain your professional demeanor. Make sure that your cell phone, pager, and all other communication devices are turned off. Also, do not chew gum, wear tinted glasses or sunglasses, or lean back in your chair.
- Dress appropriately, as you would dress for work.
- Listen to the questions before answering and always tell the truth.
- After you answer a question, stop talking.
- Be serious, courteous, and professional.
- You can step down from the stand when the judge tells you to.

## Summary

It is the revenue protection investigator's responsibility to establish liaisons with the police and fire departments and other appropriate agencies. You should also be communicating with other utilities in the service area and with utilities surrounding the area.

If you have never testified at trial, visit the courtroom for half a day. See how the criminal justice system actually works. If you are going to testify for the fist time, try to discuss your testimony with someone from the prosecutor's office or the utility attorney. Before you testify, review the witness checklist.

# Recommended Actions

- Develop liaisons with the local police and fire departments. Also get to know the codes, building, and electrical inspectors.

- If your utility has the resources, consider producing a training video for public safety agencies in your area.

- Consider an annual breakfast meeting and invite personnel from all the agencies and offices you liaison with. Keep it brief and informative. Ask for input on how the revenue protection program can be improved.

- If you are asked to participate in a major crime investigation, review that section of this chapter. Be sure that you understand your role and what the law enforcement agency expects from you.

- When preparing a case file for court, make sure that you have all the information listed in the file. The physical evidence must remain in the evidence locker until you are asked to bring it to court.

- If you have never been to court, visit the courtroom to see how the justice system actually functions.

# References

U.S. Department of Justice, Federal Bureau of Investigation. 1999. Crime-scene search. In *Handbook of Forensic Services*.

# 10
# Evolving Challenges

## The Impact of Deregulation

Deregulation of the energy industry has taken many different forms around the world. In some cases it has resulted in a greater consumer awareness of the energy bill and an increase in the cost of energy. Both can lead to increased energy theft.

In the past, most utility customers didn't think about the cost of energy unless there had been a recent rate increase. They received the bill, complained if they felt it was too high, and paid it. Where deregulation has taken place, this same customer is aware of the changes resulting because of media exposure and advertising by competing energy service providers. The outcome is that customers are aware of the cost of energy, and some feel it's too high. A few of these people will be looking for ways to save on their light bill.

Whenever there is a significant increase in utility rates, there will also be an increase in theft. Higher rates help the honest thief to justify stealing energy, especially because they know about the major energy company scandals that have occurred in recent years. "If the big guys are being dishonest, then why shouldn't the little guy get away with something?"

The increased competition in the industry has, in a few cases, resulted in less cooperation between revenue protection departments. In one case a utility from the southwest sent a team of employees to a west coast utility to study their revenue protection program. The team was supposed to spend a week with the host utility. On Wednesday morning, however, the manager of the revenue protection department at the west coast utility was told that he was to cease providing information to the team until the utility determined a way to market the information as a consulting service and determine what the fees would be. Although deregulation increases competition between utilities, revenue protection should not be affected by the new marketplace. When customers are stealing utility services, everyone loses. Revenue protection departments, even those of highly competitive utilities, should communicate and work closely together.

With deregulation, you have the Utility Distribution Company (UDC) and an Energy Services Provider (ESP). For most customers, the UDC retains the ESP. But there will be other ESPs with customers in the UDC service area. If one of those customers is stealing, who is responsible for the investigation and revenue recovery? One major state that has experienced deregulations determined that the UDC will be responsible for the investigation, operating in cooperation with the ESP.

Deregulation has resulted in a number of fraud schemes targeting utility customers. The Better Business Bureau has identified some of these:

- *Shocking*, similar to *slamming* in which some customers are switched to another energy service provider without their consent.

- Deceptive ads, such as *two free weeks*, which might include electrical costs but not fixed transmission and distribution costs, which can represent two thirds of an electric bill.

- Exorbitant hidden fees charged for folding an electric bill into a mortgage payment, or *inflation protection* with a fixed-term rate.

- Misleading information about the source of electric power, claiming it is *green* (from water, wind, or solar) rather than pollutive (from coal or nuclear sources).

- Pyramid or *Ponzi* schemes in which consumers are recruited to invest in a fly-by-night energy service provider. The profits actually come from other investors until the scheme collapses.

- Online investment schemes with fabricated press releases, impressive looking charts, and so called *insider information* about what is in fact a bogus deal.

In one pyramid scheme, a 19-year-old college student in Pennsylvania, Christopher S. Mee, sold electric distributorships to customers in California and other states using the company name Boston-Finney, Inc. "You too can be an ESP." Almost 8,000 people signed up, paying $300 to become *power reps*, plus $160 annual dues. (See Case No. 717631, Calif. Superior Court, San Diego County; Docket No. 183 MD 1998, Pa. Commonwealth Court.)

As a result of an investigation into the case, a judge in Pennsylvania issued an injunction, freezing $310,000 in personal and business accounts and impounding Mee's $90,000 Mercedes-Benz, bought with corporate funds. A year later he apologized to the court for violating previous court orders aimed at stopping the pyramid scheme and was ordered to pay almost $300,000 in fines, back rent, and restitution. The proceeds from the sale of the Mercedes-Benz were placed in a fund to compensate victims. While it was in operation, Boston-Finney collected as much as $2.36 million. Mee was not sentenced to any jail time.

Another scheme targeting utility customers in deregulated areas is the bundling of services. The customer receives a significant discount on the electric bill if he also signs up for local and long-distance telephone service and other services. The prices for these services are inflated to make up for what the provider is loosing by providing electricity.

## Changes In Metering

Advanced technology is changing the way utilities meter customer usage. Mechanical meters are being replaced with solid state electronic meters and automatic metering systems (AMRs). The results of a 2002 utilities survey found that 20% of utilities use or plan to use AMR, 35% were considering using AMR for tamper detection, and tamper detection was the most popular benefit of AMR. Most of these systems include features to deter and detect tampering. However, the determined thief will still get you. They just have to be more clever than in the past.

Electronic meters alone still require meter readers. With some of these products, the meter reader does not have to actually see the meter but can take the reading into a handheld receiver using a radio read or other technology. This is convenient for the meter reader, and it may be convenient for the customer, too. Meters that are not read each month are a temptation for some customers.

Electronic meters are programmed and can be reprogrammed. They require user identification, including passwords, to protect the programming. If the customer could access the program, he could change the output—the meter reading. The average customer would not have the capability to hack into an electronic meter, but the threat must be considered.

How many protected computer systems have been hacked by teenagers? Hackers have been able to break into Defense Department computers, the White House servers, the Federal Bureau of Investigation (FBI) and more networks than we will ever know. A single electronic meter may be the next challenge. One of the more popular hacker publications has been running articles on energy theft since 1995. An April 1996 issue cover featured three electric meters, two mechanical meters, and an electronic meter with the caption reading,

> Hey, does this look familiar? I'm sure we've talked about these before…right? Anyhow, as you can see, we've always had "easy access" to power meters. Do you have any idea what this means and how unsecure the entire electric provider industry is? Ok, so the power companies put these little security locks and tamperproof tags on them. BIG DEAL! Theft of electricity is at its highest point ever and it keeps getting worse.

You don't think hackers are looking at electronic meters? Think again. There is a new web site on the Internet that states, "Find the best sites for electric meter hack with XXXXware. XXXXware search is an excellent resource for quality sites on electric meter hack and much more. XXXXware also provides related listings for electric meter hack."

Once a utility has enough electronic meters in the field to make it worthwhile, the new fixer will be a hacker with a laptop. The electronic hacker will use password cracker software to find the system's password and then reprogram meters. This is easier than hacking computer systems because each computer system has a different password. With electronic meters the utility often uses the same password for all of their meters. Once the password on one meter is cracked, there is unlimited meter password access.

Many utilities are turning to AMRs to solve metering problems they face in the new world of competition and cost control. There are many advantages to AMRs:

- Precise load forecasting
- Reduced maintenance costs
- Advanced outage detection
- Improved system reliability
- Proactive load management
- Interruptible rates
- Two-way control
- Accurate, timely billing
- Improved customer service
- Unnecessary to access the customer's property
- Selectable billing dates
- Monitoring services

There are many advantages to AMR systems, but there are also some potential challenges. A utility with more than a million customers is going 100% AMR, predicting that energy theft will disappear as the system is implemented. Instead, theft is increasing. Customers have learned to put a bypass around the system during outages. Using this method none of the tampering alarms on the system are triggered. And because meter readers are no longer looking at the system each month, the bypass does not have to be as well concealed as in the past.

Another utility is testing an AMR commercial metering system that uses cell phone frequencies. Customers continue to steal meters from inactive services, but when they put the meter in the its base at the thief's location, it sends a signal back to the utility that shows the meter is energized and provides a meter reading. The problem is that the utility doesn't know where the stolen meter is. Because they are using cell phone technology, they can trace the signal to the cell phone tower that is picking up the signal and limit their search to that area.

Some of the ways customers could steal from AMR systems include the following:

- AMR module tampering
- AMR message tampering
- Cloning the AMR module
- MDMA database tampering

The thief could interfere with the AMR consumption sensor or local communication port. By interfering with the sensor reliability, the recorded usage could be reduced. The customer could also alter the data within the module and reduce the billing usage.

It is possible to block the communication from the AMR and replace actual messages with alternative data. This requires considerable expertise, but it could be done. Rather than tampering with the AMR message, the thief could use an AMR module duplicate, possibly one stolen from another user with lower consumption. The customer inactivates his meter and sends the utility the reading from the cloned or stolen meter.

Meter Data Managing Agents (MDMAs) are contracted companies. They supply meter reading data to UDCs and ESPs. The threat here is that a customer could be working in collusion with an employee who modifies the metering data for a percentage of what the customer saves.

Most of the AMR systems marketed today have programs to detect each of the tampering schemes discussed as well as traditional methods such as magnets on the meter or inverting the meter. The major concern with AMR is that the utility is not looking at the meter each month so the theft could go on for some time without detection.

From a proactive perspective, AMR allows the utility to monitor the customer's consumption frequently. The system can be programmed to detect unusual changes that warrant investigation in usage on an hourly or daily basis, rather than every 30 days.

The ultimate threat to AMR systems is a denial of service attack (DOS). A DOS attack could shut down the entire system and result in the loss of data that had not been backed up. Hackers love denial of service attacks.

There are three basic steps to protecting AMR systems with regard to revenue protection. First, select the best system and application for your area and objectives, and install the system according to vender specifications. Next, ensure that the communication system is secure and that all information technology (IT) protocols are in place. Finally, visually inspect every AMR meter at least once a year; every six months is better. There is no substitute for having a utility employee physically inspect a meter.

## Sarbanes-Oxley Act of 2002

One of the newest factors impacting utilities owned by publicly held companies is the Sarbanes-Oxley Act of 2002. Triggered by major corporate financial scandals, the Sarbanes-Oxley Act requires companies under the act to

- Review current business practices, which have an impact on the financial statement and develop a flowchart of these activities.
- Identify the key risks associated with each activity in the process.
- Prepare a control matrix to prevent and detect risks.

Complying with Sarbanes-Oxley can be a challenge, but over all it should be an asset to revenue protection. Failure to have an effective revenue protection program leads to several major risks including the loss of significant revenue and danger to customers. As discussed in chapter 1, revenue protection is both an economic and risk management issue.

Utilities covered by Sarbanes-Oxley are forced to examine and flowchart revenue protection activities. This should result in a more efficient program. When the financial impact of the program is examined, it should illustrate that the revenue protection program identifies and recovers revenues that would have been lost.

If your utility must meet the requirements of Sarbanes-Oxley, revenue protection can expect to spend more time with accounting and IT personnel, as well as with external consultants. Additional internal controls will be developed to monitor the performance of the revenue protection program.

# Protecting Investigators and Investigations

The revenue protection investigator must be prepared to face threats from angry customers and employees fired for theft. The investigator must also protect the information collected during the investigation. If there is an ongoing conspiracy investigation, someone may attempt to compromise the information during the investigations, especially if professional criminals are involved.

If you are conducting an internal investigation, the suspect employee may attempt to gain access to investigative information. For example, a utility's security unit had purchased a covert surveillance system to place at certain locations to detect possible employee theft. They decided to test the equipment, which had motion detection activation, by leaving it in the office overnight to see if it recorded the cleaning crew's activities. It not only recorded the cleaning crew, but it also recorded an employee under suspicion for meter tampering coming into the office and checking the file on his case to see if there was anything new. The recording of the employee's actions helped build the case against him. The fact that the employee could not only get into the security office but could also get into their files doesn't speak well of the security office's security.

There are three areas of security that the revenue protection department should consider:

- Operations security
- Physical security
- Personal security

According to one definition

*Operations Security (OPSEC) is a risk management tool used to deny an adversary (The Bad Guys) information concerning our intentions and capabilities by identifying, controlling, and protecting indicators associated with the planning and execution of the organization's mission.*

In terms of revenue protection, this means not giving the customer the information needed to easily steal energy and not giving the thief the information needed to get away with stealing energy. To develop an OPSEC program, the revenue protection department needs to identify the Essential Elements of Information (EEI) the bad guys need to get away with stealing services and then developing programs that prevent them from obtaining that information. Begin by considering the EEI that a thief would like to have:

- What day of the month will my meter be read? *Make sure the service is normal on that day.*

- Does the utility use an exceptions report to identify meters to be inspected? *Steal less than the exceptions report is looking for.*

- Who are the revenue protection investigators? *What vehicles do they drive?*

- Does the utility check meter seals on a regular basis? *Can I obtain extra seals?*

- If a utility catches a customer stealing, what are the consequences?

The OPSEC program is developed during three brainstorming sessions. At the first session develop a list of EEI. If you were planning to steal energy, what information would you like to have? At the second session review the first list and determine how the customer would obtain this information. At the final session, brainstorm methods that could be used to prevent the customer from obtaining the EEI.

The revenue protection office must be secured. In addition, you must consider the security of the evidence stored there and of your investigative information. The office should have secure locks that cannot be opened by widely distributed keys. The door should be locked even when investigators or staff personnel are in the office. You may want to put a keypad lock on the door for additional security. Consider leaving the trash outside of the office at night so that cleaning personnel need not access the area. When conducting revenue recovery interviews, use a meeting room outside of the revenue protection office. Keep this area secure at all times.

Evidence must be locked inside a container in the office. The container can be a closet or a file cabinet. It is locked at all times and only the investigators should have keys to the evidence containers.

What about the information on your investigations? If this information is stored as hard copies, then it should be in a locked file cabinet. If you are storing investigative information on a computer, password protect the computer or at least the files with the information. If it is stored on a network, talk to your network administrator. There are a number of ways to secure the information on the network, from user identification to encryption. If you are really suspicious, use both.

Some approaches to personal security were discussed in chapter 8. As a revenue protection investigator, you are going to make some enemies. Here are some additional tips to improve your personal security:

- Do not put personalized license plates or bumper stickers on your personal vehicles.

- Consider an unlisted telephone number at home. If this is impractical, ask the telephone company not to list your street address in the telephone book.

- Conduct a personal Internet search. Enter your name on your favorite search engine and see if there is information available. Is your home address available on the Internet?

- Keep a low profile.

- Don't give out personal information on the telephone. Instruct family members to do the same.

- Be alert to strangers around the house and in or near the workplace.

- Vary your routes and times to and from work and home.

You are a revenue protection investigator, so take the personal threat seriously. Remember the old saying, "You're not paranoid if someone is really after you."

## Summary

Almost every utility should have a revenue protection program. In small systems revenue protection responsibilities are assigned as additional duties to employees in other positions. In larger utilities there should be full-time revenue protection personnel. In either case these employee must be trained to conduct investigations and to recover lost revenues. If the utility does not have a revenue protection program, then the losses are built into the rate base. This isn't fair to the honest customers.

## Recommended Actions

- If your utility has had a rate increase recently, then you may also have experienced an increase in energy theft. Have you intensified your efforts to detect tampering, diversion, and fraud?

- Electronic meters and AMRs have features to deter and detect energy theft. If your utility is using these meters or systems, discuss these features with the metering department and the manufacturer's representative. Identify the potential vulnerabilities of the systems and their uses as well. That's what some of your customers are doing.

- If you are using AMR systems, how frequently do you visually inspect meters? It should be at least every six months, although some utilities schedule annual inspections. Regardless of the frequency, randomize the inspections to that the customer cannot predict when they will be conducted.

- Review the security procedures in place to protect investigations and investigators. Law enforcement officers can provide additional suggestions, especially regarding personal security. Develop an operations security program to protect information on your procedures and investigations.

- Conduct a physical security survey of your office area. This is mostly common sense, but a crime prevention officer could provide additional suggestions.

- If your investigative information is stored on a network, talk to the system administrator to make certain that it is protected from unauthorized access.

# References

*Blacklisted! 411, The Official Hackers Magazine.* April 1996. 3(2).

DeKok, D. 1998. Boston power dealer to pay $300,000 in pyramid scheme. *Knight Ridder/Tribune Business News,* June 17, 1998. http://www.highbeam.com/library/doc3asp?ctr1Info=Round5%AProd%3ADOC%3APrin (accessed July 29, 2004).

Electric Meter Hack. Find the best sites for electric meter hack with Starware Energyshop. What to watch for. http://www.energyshop.com/energyshop/e-watch.cfm (accessed July 29, 2004).

Soyka, D. Pyramid Schemes: A Black Eye for Power Retailing? *Public Utitlities Fortnightly.* May 1, 1998, http://www.highbeam.com/library/doc3.asp?DOCID=1G1:20587137&num=3&ctrlInfo=R (accessed July 29, 2004).

Malemezian, E. 2003. Revenue assurance with AMR. The good, the bad, and the possibilities. *International Utilities Revenue Protection Association* 14(2).

Soyka, D. May 1, 1998. Pyramid schemes: A black eye for power retailing. *Public Utilities Fortnightly.*

Weibe, M. 2000. *A Guide to Utility Automation: AMR, and IT Systems for Electric Power.* Tulsa, OK: PennWell 2000.

Woodward, W. 1997. Revenue protection in a deregulated environment. *International Utility Revenue Protection Association Newsletter*, December 12. http://www.iurpa.org/ed-dereg.htm (accessed July 28, 2004).

# Position Description

**Position:** Utilities Service Investigator

**Dot Number:** 821.365.010

**Primary Responsibilities:**

Investigates reports of METER READERS (utilities, waterworks) concerning evidence of tampering with watt-hour meters and related wiring that interferes with correct metering of electric current consumed. Requests permission of customer to examine meter. Examines meter for evidence of tampering. Examines wiring connecting meter to outside power lines and building circuits. Removes unauthorized wires or devices illegally installed to bypass current past meter, or to obstruct normal meter function. Turns on bulb of known wattage and observes movement of meter disk, determining accuracy of disk rotation in relation to current consumed by the bulb. Replaces blown fuses and repairs minor defects, such as loose wiring connections or insulators. Removes meters showing evidence of tampering and informs customer of discontinuance of service, pending settlement with company. Compiles inspection report, confirming or denying suspected tampering, and reports needed major repairs. May serve court summons and testify in court as company witness during criminal prosecution.

## Scope of Responsibility:

Responsible for planning and completing assignments under limited supervision. Assignments tend to be individual in nature and require limited input from others within the work area. May be responsible for supervising a defined work group to accomplish a specific task.

## Duties and Responsibilities:

1. Investigates cases involving meter tampering, fraud, or other means of energy theft.

2. Receives information regarding suspected theft or fraud involving electric services and equipment. Coordinates with other departments to expedite cases from initial discovery to conclusion.

3. Interviews employees and observes and questions customers to obtain evidence. Collects and preserves evidence to prepare cases for court.

4. Prepares case summaries and testifies at hearings of suspects and/or coordinates the negotiation for restitution with customers or customer's representative.

5. Prepares and disseminates training materials to employees and the public regarding theft of service and the revenue protection program.

## Education and Experience

Position requires a Bachelor's degree; two years of related experience; and a working knowledge of metering systems.

## Knowledge and Skill

Position requires an in-depth knowledge of technical and clerical fields. Must understand the policies and procedures of the job to handle unusual situations.

## Special Licensing and Certification

Valid driver's license.

## Working Conditions

The working environment for this position includes physical conditions that may be undesirable. The environment may include noise, dust, heat, oil, dampness, or weather. The job demands light physical effort. May require walking or standing for long periods of time. Requires prolonged machine operation including keyboard equipment.

# References

U.S. Department of Labor, Employment and Training Division. 1991 (rev). *Dictionary of Occupational Titles, Volume II, Fourth Edition.* Washington, DC: U.S. Department of Labor.

CITY OF _____ ELECTRIC DEPARTMENT

PROCEDURES TO HANDLE

UTILITY THEFT AND FRAUD SITUATIONS

FOUND IN THE FIELD BY SERVICE

PERSONNEL

**Revenue Protection:** Combating Utility Theft & Fraud

## CONTENTS

| TOPIC | | PAGE |
|---|---|---|
| Section I. | Most Common Violations and Procedures When Handling Possible Theft and Fraud Situations | 183 |
| Section II. | Illegal Reconnect on an Inactive Account with Reconnect Order When Less Than 100 kWh Has Been Registered Since the Last Disconnect | 184 |
| Section III. | Cut Seals | 184 |
| Section IV. | Extension Cords | 185 |
| Section V. | Damaged Meter | 185 |
| Section VI. | Damaged or Removed Locking Band or Lock—not resulting in the illegal use of power | 186 |
| Section VII. | Yellow Seals | 186 |
| Section VIII. | Fraudulent Service | 186 |
| Section IX. | Danger to Life, Limb, or Property | 187 |
| Section X. | Best Judgment | 187 |

Appendix B

## I. Most common violations and procedures when handling possible theft and fraud situations

The following is a list of the most common utility theft violations found in the field, followed by the proper procedures for handling these situations.

(A) Jumpers on the rear of the meter or in the meter socket.

(B) Illegal connection at the weatherhead.

(C) Illegal connection in an inactive account with reconnect order if over 100kWh have been registered since the disconnect.

(D) Illegal reconnect on a nonpay account.

(E) Illegal reconnect on a nonpay account with a reconnect order.

(F) Meter bypass.

(G) Meter upside down.

(H) Damaged or removed seal, lock, or locking band.

(I) Damaged meter found at location that has been disconnected for nonpayment, utility theft, returned check, or service disconnect.

(J) Illegal connection on inactive account.

(K) Use of stolen meter or any other unauthorized meter at an inactive or active account.

(L) Fraudulent use of name on account to obtain service.

**Meter Readers and Service Personnel:** On discovering any of the previously mentioned conditions follow the procedure described:

(A) If the situation is a *suspect* theft or fraud (for example, a cut seal), report the situation to your supervisor or to the Codes Enforcement Officer at the end of the work day.

(B) If the situation is obvious tampering (for example, meter inverted), report to the dispatcher immediately and continue reading meters or working work orders.

(C) If the situation is dangerous (for example, an open meter socket), report to the dispatcher immediately and wait at the location until a service person arrives to render the service safe. If the customer is present, tell the customer that there is an unsafe condition and that a crew is en route to correct it.

**Dispatcher**: When informed by meters readers or other field personnel of an obvious theft or an unsafe condition, notify the Codes Enforcement Officer and appropriate service personnel immediately.

## II. Illegal Reconnect on an Inactive Account with Reconnect Order When Less Than 100 kWh Has Been Registered Since the Last Disconnect

**Service Personnel:** On discovering usage of *less than 100 kWh* on an inactive account that has a connect or reconnect order, record the meter reading and the findings on your work order and connect the service unless there is damage or a yellow seal on the meter. If you find damage or a yellow seal, do not connect or reconnect the service.

### III: Cut Seals

**Service Personnel:** Report all cut or missing seals to the dispatcher.

**Dispatcher:** Forward all reports of cut or missing seals to the Codes Enforcement Officer.

## IV. Extension Cords

When extension cords are used to provide power to a location without service, the service at the active account will be terminated immediately. Both customers must contact the Codes Enforcement Officer before service can be reconnected at either account. The use of extension cords in these cases creates a potentially dangerous situation that could result in fire and loss of life.

> **Service Personnel:** On finding extension cords used to provide service to an inactive account, immediately report the condition to the dispatcher. Remain at the location until the service at both locations is disconnected.
>
> **Dispatcher:** When notified of extension cords being used to provide service to an inactive account, immediately dispatch a crew to disconnect both services and notify the Codes Enforcement Officer.

## V. Damaged Meter

A. Trouble Call

> **Service Personnel:** On finding a damaged meter when dispatched to a location because of a reported service problem, contact the dispatcher before correcting the situation and restoring service.
>
> **Dispatcher:** On receiving a call from the service person, ensure that the service is not disconnected as a result of nonpayment, utility theft or for other reasons and that it is an active account. Unless one of these reasons exists, instruct the service person to correct the situation and restore service.

B. Meter Accidentally Damaged by Service Personnel

> **Service Personnel:** If the meter is damaged by personnel working on the service, restore the service and make appropriate meter change orders. Record what occurred. No disciplinary action will be taken, but it may be necessary to review the use of equipment or the procedures followed or the need for additional training.

### VI. Damaged or Removed Locking Band or Lock—Not Resulting in Illegal Use of Power

**Service Personnel:** Report the situation to the dispatcher immediately. Leave the situation as found. If the situation is not dangerous, continue reading meters or working work orders.

**Dispatcher:** Report immediately to Codes Enforcement Officer.

### VII. Yellow Seals

**Service Personnel:** Yellow seals are used only on accounts where tampering has been detected or where it is suspected. Yellow seals should not be removed under any circumstances unless authorized by the Codes Enforcement Officer, the operations manager, or the general manager. If a yellow seal is found cut or otherwise damaged, notify the dispatcher immediately.

**Dispatcher:** When receiving a report of a cut or damaged yellow seal, notify the Codes Enforcement Officer immediately.

### VIII. Fraudulent Service

Some customers attempt to obtain fraudulent service by using incorrect names on the service applications. This is especially prevalent where service has recently been disconnected for nonpayment.

**Field Personnel:** If you receive a connect order for a service that you believe may be fraudulent, do not connect the service. Notify the dispatcher of your suspicion and the reasons for being suspicious.

**Dispatcher:** On receiving the report from the field personnel, notify the customer service manager and the Codes Enforcement Officer. The customer service manager will review the records on the account to determine is fraud may be occurring. The Codes Enforcement Officer will inspect the service; service should not be connected until approved by the customer service manager.

Customer Service Manager: If it is determined that fraud may be occurring at the account, the customer will be required to come to the service center to verify his or her identity and to personally attest that a disconnected customer is not living at the residence. If fraud is occurring or has been attempted, the customer will be required to pay all back bills and current charges before service is connected. If it cannot be determined that fraud is occurring or has been attempted, the service will be connected.

## IX. Danger to Life, Limb, or Property

Under no circumstances are these procedures intended to over-ride or supersede the obligation of _____ to prevent, if possible, the injury or loss of life, limb, or property. Safety should always be the first consideration.

## X. Best Judgment

These procedures cannot cover all the situations involving tampering and fraud that employees find in the field. These procedures provide guidelines that should be used in combination with the employee's knowledge and experience. When in doubt about how to respond to a unique situation, contact the dispatcher. The dispatcher will then contact the appropriate personnel to assist you.

# Draft Ordinance

ORDINANCE NO._____

AN ORDINANCE PROHIBITING CERTAIN PRACTICES AND PROCEDURES AND THE ILLEGAL RESTORATION OR OBTAINING ELECTRICAL CURRENT BY OTHER FRAUDULENT MEANS; DECLARING AN EMERGENCY AND FOR OTHER PURPOSES.

WHEREAS, the incidents of electric theft from the _____ Electric Department have greatly increased, and

WHEREAS, the City Council of the City of _____ determined that certain practices must be prohibited to curb these practices.

NOW, THEREFORE, BE IT ORDAINED BY THE CITY COUNCIL OF THE _____:

SECTION 1: That the following practices or acts are hereby declared to be illegal and are prohibited by the _____ City Council:

1. The cutting, removing or in any manner making ineffective any seal, locking band or lock on an electric meter.

2. Restoration of service by any means after service has been terminated for nonpayment or obtaining electricity without making the

proper deposit with the _____ Electric Department or receiving proper authorization from the _____ Utilities Accounting Department or the _____ Electric Department.

3. Obtaining electricity by use of jumper wires, or by any other means which bypass a metering device either partially or completely.

4. Damaging or tampering in any manner with any part of an electric metering device belonging to the City of _____.

5. Changing or altering the normal installed position of a metering device in any fashion, which causes the normal accurate recording of utility service received to be altered.

6. Obtaining electricity by use of a metering device, which is not authorized and installed by the _____ Electric Department.

7. Interfering with the automatic registration, recording, and transmission of electricity consumption when readings are recorded and/or transmitted electronically.

8. Fraudulently obtaining, or attempting to obtain, service from the _____ Electric Department by using a false name or identification.

9. Fraudulently obtaining, or attempting to obtain, service from the _____ Electric Department by placing the account in the name of some else after the service has been disconnected for nonpayment or theft of service while the occupant who was originally disconnected is still living at the location of the service.

10. Any other reception of electrical power without proper authorization from the _____ Electric Department or the _____ Utilities Accounting Department.

SECTION 2: Any person found guilty of violating any section or part of this ordinance shall be fined a sum not less than $100.00 or more than $500.00 on each count.

SECTION 3: Any person found violating any section or part of this ordinance will be required to make restitution to the _____ Electric Department for the cost of the electrical services obtained in violation of the ordinance, for damage to _____ Electric Department's equipment, and for the cost of the investigation.

1. The cost of the service received as a result of the violation of this ordinance shall be computed by the _____ Electric Department.
2. The cost of any damage to utility equipment shall be based on the actual costs. These costs may be computed and updated annually and submitted to the City Council for approval.
3. Investigative costs will be computed and updated annually and submitted to the City Council for approval.

SECTION 4: Any person in possession of any premises as owner, occupant, or tenant who is found to have an electric meter that has been tampered with or altered in violation of this ordinance or found to be receiving electricity as a result of any of the methods described above shall be presumed to have knowingly violated the terms of this ordinance. Such receipt of electricity shall be deemed prima facie evidence of intent to defraud or deprive the _____ Electric Department of recovering proper charges for payment for such electric service.

SECTION 5: All ordinances or parts of ordinances in conflict herewith are hereby repealed to the extent of the conflict.

SECTION 6: That the provisions of this ordinance are hereby declared to be severable; and if any section, phrase, or provision shall be declared or held invalid, such invalidity shall not affect the remainder of the sections, phrases, or provisions.

SECTION 7: The City Council of the _____ finds that there is a continuing problem with persons illegally obtaining electricity by various means and that this results in a loss of revenue to the City and creates a dangerous situation that can result in the loss of life or property; THEREFORE, an emergency is hereby declared to exist and this ordinance being necessary for the preservation of the public peace, health, and safety shall be in full force and effect from and after its passage and approval.

# Twenty-fourth Guam Legislature

1997 (First) Regular Session

AN ACT TO CLASSIFY UNAUTHORIZED ELECTRICAL CONNECTIONS OR "ILLEGAL HOOKUPS" AS A THEFT CONSTITUTING A FELONY OF THE THIRD DEGREE AND TO ESTABLISH FINES AND PENALTIES FOR SUCH HOOKUPS

BE IT ENACTED ON BY THE PEOPLE OF THE TERRITORY OF GUAM:

**Section 1.** The legislature finds that, according to the Guam Power Authority, there is an increasing number of unauthorized electrical connections or "illegal hookups" on GPA power lines. Theft of electrical current is a growing problem, and is at a point where something has to be done. The legislature finds that during the past year, the dismal state in the unreliability of the power system on Guam constituted a threat to the health and safety of our people. There is no doubt that the people of Guam have suffered tremendously from the incessant breakdowns and failures of the power generating system. In addressing this plight, the people of Guam have banded together commendably. In facing this predicament of powerlessness, many have come to tacitly accept the rate increases by the Guam Power Authority, have come to accept the state of emergency declared by the governor in order to procure power generation from private sources, and have come to adapt remarkably to numerous power outages of varying time lengths that come largely unannounced throughout the day and night. The legislature therefore finds that unauthorized electrical connections or "illegal hookups" by individuals, contractors, or businesses seeking to avoid paying their fair share of electrical power, is something not to be taken lightly. The legislature finds that more stringent laws are needed in order to help prevent these illegal hookups, and to prevent a useless and illegal drain on our island's available wattage.

**Public Law #24-31**

**Bill 3426**—An Act to amend PL#24-31 relative to classifying unauthorized electrical connections as theft, to adopt proposed fines and penalties for such hookups. Passed by the 24th Guam Legislature on February 5, 1998. Approved and signed by the Governor into Law on February 16, 1998.

**Section 5. Education and Amnesty Period.**
This section shall become effective 45 days after the enactment of this Act to allow for an educational period in which GPA shall conduct a public awareness campaign on the provisions of this Act.

There shall be an amnesty period of 90 days beginning after the expiration date of the 45 days educational period for users wishing for assistance in disconnection and/or correction of conditions, which are in violation of Section 2 of this Act. Once a customer has reported to GPA that they are in violation of this Act and requests assistance in correcting the conditions, it is the responsibility of GPA to correct the conditions without penalty to the customer.

## Section 6. Establishment of Reward for Persons Reporting Illegal Power Hookups.

A reward of $500.00 shall be paid by GPA to any person who reports an illegal power hookup resulting in collection of fines and penalties, or conviction. The name(s) of any person reporting illegal power hookups shall remain confidential.

## Section 7. Fines or Penalties for Unauthorized Electrical Connections.

Every user, contractor, or business who knowingly taps into, tampers with, alters, or by-passes electrical meters in accordance with the provisions set forth in Section 2 of PL #24-31, shall be subject to penalties that will be calculated at twice the amount of the estimated value of the power utilized by the illegal connection. Said penalties are exclusive of the estimated value of the power utilized by the illegal connection that is due and payable to GPA in accordance with the provisions set forth pertaining to the back billing.

A payment schedule for a period of 24 months, in which imposed fines shall be paid along with the regular power bills, at 12% annual interest on the unpaid balance.

Any user who makes an unauthorized connection shall have GPA service discontinued and shall be refused service until such time that the aggregate amount of penalties are paid or a payment schedule has been arranged and approved by GPA.

The General Manager of the Guam Power Authority shall formulate a penalty schedule based on $2,500.00 for illegal hookup including meter tampering, up to $25,000.00 maximum. The General Manager may formulate a payment schedule for the penalty up to 24 months, to be paid along with regular power bills at 12 percent annual interest on the unpaid balance. Any person who makes an unauthorized connection shall be refused or discontinued service until the penalty is paid. The owner and customer on record, if any, of any property having an illegal power hookup or where meter tampering is found, shall be personally jointly and severally liable for all penalties and estimated value of the power used.

# Revenue Protection Sample Web Page

**Electric Theft** is an international problem. Billions of dollars are stolen from utilities each year. These costs are often passed onto utility customers in the form of higher rates. Your Friendly Electric Utility has taken a proactive stance on electric theft. It is our goal to stop electric theft and keep your rates from increasing to pay for what dishonest customers are stealing.

**Electric Theft** is also a public safety issue. Many cases that are discovered turn out to be electrocution and fire hazards for the thief and others in the neighborhood. These conditions can lead to property damage, personal injury, and even death.

**Electricity Theft** is a crime. Electric thieves can go to jail.

Please contact your Friendly Utility to report electric theft. Your anonymous tip will result in an immediate investigation and could result in the recovery of lost revenues. It may also prevent a fire, injury, or death.

# Revenue Protection Investigation Report

Investigator's Name:_____

Date:_____ Time:_____

Customer of Record Name:_____

Customer's Address:_____
_____

Customer #:_____ Service Loc #:_____

Meter #:_____ Seal #:_____

Last Meter Reading:_____ Current Reading:_____

1. Date and time that you arrived at the location.
2. Why were you at this location?
3. Is this an active or inactive service? If inactive explain; disconnect for nonpay or no current application for service at this account.
4. What did you observe? (Example: condition of seal and meter.)
5. Were other employees present during the inspection?
6. Was the customer present during the inspection? If so, describe discussion.

7. What did you do? (Example: remove meter, check disk, check meter cover.)

8. Did you photograph the inspection scene? (Include photo log with report.)

9. Was there any damage to utility property? If so, describe.

10. Was service disconnected?

11. Was evidence secured while in the field? Describe where and how secured.

12. Was evidence secured in the office? Describe where and how secured.

13. Have there been previous investigations of this customer or this location? If so, describe.

14. Attach copy of meter test results to report.

15. Attach copy of back bill.

16. Attach copy of the service agreement, if available, to this report.

17. Was case referred to law enforcement? If so, discuss reasons and agency referred to.

Document all contacts with the customer regarding this case.

# Index

1800ussearch.com, 79

## A

Account closed check, 63
Accounts not consuming, 37
Activity report (revenue protection program), 52
Advanced techniques (investigative interview), 107–109
Agricultural theft, 10–12, 76–77
Altered check, 63
Alternative question (interview), 108
American Society for Testing and Materials, 74
Amnesty period/program, 24–26:
 employees, 24;
 customers, 25
Anatomical/physical responses (interview), 104
Anger and aggression (dangerous accounts investigation), 113–128:
 the problem, 113–115;
 drug manufacturers, 115–120;
 indoor marijuana growing, 120–123;
 defusing, 123–126;
 escalation factors, 124–125;
 summary, 126;
 recommended actions, 127;
 references, 128

Anger escalation factors, 124–125
Anti-tampering devices, 28
Apology (anger), 124
Appearance (trial), 159–160
Application for service fraud, 59, 62:
 residential, 59;
 commercial, 59;
 combating, 62
Arizona Public Service Company (APS) study, 6–7
Assessment questions, 94–95, 107
Attorney fees, 81
Attorney presence, 98
Australia, 3, 114
Automatic metering system (AMR), 167–171, 175
Awareness campaign (public), 18–19, 26:
 element, 18

## B

B.C. Bud (cannabis), 13
Back billing (revenue recovery), 21–23, 80–86, 89:
 collections, 21–23;
 computing, 80–86, 89;
 records, 83–84;
 billing policy, 89

Bad checks (fraud), 59–60, 62–64, 68:
    residential, 59;
    commercial, 60;
    combating, 62–64
Bankruptcy fraud, 64–65
Bankruptcy records, 75
Basic premise (case), 156
Behavior (detecting deception), 102–107, 109:
    pronouns, 106;
    nouns, 106;
    verbs, 106;
    repeated accounts, 107;
    responses, 109
Belgium, 116
Best judgment (procedures), 187
Blame (anger), 125
Blue seal account, 27
Booby traps, 123
Braddock's Federal-State-Local Government Directory, 72
Bundling of services (fraud scheme), 167
Business information services, 75
Business information sources, 73
Business legal structure, 65, 68
Business sources (information), 71, 73
Bypassed meter, 1–2, 13, 183

## C

California Commission on Peace Officer Standards and Training, 46–47
California, 46–47, 115
Canada, 6, 8, 12–13, 120–123:
    Operation York Connection, 121
Case assignment priorities, 41
Case file preparation, 158–162
Case histories, 1–6, 113–115:
    dangerous accounts, 113–115
Case issues/elements coverage, 100–101
Case review/pattern recognition, 15
Case supervision, 41:
    assignment priorities, 41

Certification/licensing (investigator), 178
Challenge and tools (investigative), 138–145
Challenges evolving, 165–176:
    impact of deregulation, 165–167;
    changes in metering, 167–171;
    Sarbanes-Oxley Act of 2002, 171–172;
    protecting investigators and investigations, 172–175;
    summary, 175;
    recommended actions, 175–176;
    references, 176
Change-out meters, 37
Character testimony, 103
Charges and fees, 89
Check kiting, 63
Check meter, 82–83
Check washing, 64
China, 4
Chronological order (documentation), 49–50
Circumstantial evidence, 42
City ordinance (draft), 189–192
Civil legal proceedings, 86, 88
Civil litigation records, 74
Closed account check, 68
Closed questions, 94, 99
Clues of deception, 102–105:
    verbal, 102–104;
    nonverbal, 104–105
Cocaine, 13, 117
Code inspectors, 149, 152, 184–187:
    Codes Enforcement Officer, 184–187
Collecting evidence, 31, 39–40, 43–45, 50
Collecting lost revenue, 21–23, 86–88
Collective bargaining agreement (labor union), 131–132, 146
Combating utility customer fraud, 62–68:
    application for service, 62;
    bad checks, 62–64;
    planned bankruptcy, 64–65;
    out of business/back in business, 65;
    skips, 66

# Index

Commercial accounts, 10–11, 37–38, 59–60, 62, 68, 76–77, 88:
theft of service, 10–11, 59–60, 76–77;
fraud problems, 59–60

Commercial and industrial cases (information sources), 76–77

Commercial fraud problems, 59–60:
fraudulent application for service, 59;
theft of service, 60;
skips, 60;
planned bankruptcy, 60;
bad checks, 60;
out of business/back in business, 60

Common violations (procedures), 183–184

Communication skills, 108, 126–127

Communications (utility), 15, 67, 153, 165–167

Compare usage method, 85–86

Computer department, 19, 21, 23, 36, 135

Computer reports, 135

Concealing lie, 102

Confession (interview), 108–109

Conspiracy (energy theft), 135–146:
indicators/scenarios, 135–138;
challenge and tools, 138–145;
software, 145

Consumption drop, 83–84

Coordination (revenue protection program), 21–23

Corporation and assumed name records, 75

*Corpus Juris Secundum*, 73

Counterfeit check, 63

Court component (revenue protection program), 17, 20–21

Court trial, 61, 66, 149–163:
establishing liaisons, 149–153;
solvability factors, 153–155;
major crime scene investigation, 155–157;
preparing for/going to trial, 158–162;
summary, 162–163;
recommended actions, 163;
references, 163

Courthouse information, 72

Covert metering, 82–83

Crack/freebase cocaine, 113, 117

Crack house, 113, 117

Credit and collections department, 38

Credit check, 56, 67–68, 75

Credit information, 75:
sources, 75

Crime evidence, 41–45:
direct, 42;
real, 42;
circumstantial, 42;
opinion (expert testimony), 43;
documentary, 43;
scientific, 43

Crime scene investigation (major crime), 155–157:
basic premise, 156;
preparation, 156;
approach, 156;
secure and protect, 156–157;
preliminary survey, 157;
physical evidence, 157;
narrative report, 157;
photographs, 157;
sketches, 157;
final survey, 157

Crime scene investigation (theft of service), 30, 35–53, 155–157:
examination, 30;
detecting theft, 35–38;
principles of investigation, 39–41;
search, 39–40, 45–49, 155;
processing, 39–40, 45–50;
documenting, 39–40, 45–52, 157;
rules of evidence, 41–45;
recognition, 45;
summary, 52;
recommended actions, 52–53;
references, 53

Crime scene search, 39–40, 45–49, 155:
protocols, 45–46;
stages, 47–48;
procedures, 47–49

Crime scene security, 30, 156–157, 172–176

Criminal activities, 113–123:
the problem, 113–115;
drug manufacturers, 115–120;
indoor marijuana growing, 120–123

Criminal histories, 75

Criminal intelligence, 132–135, 146:
  intelligence capability, 146

Criminal justice system (trial), 158–163

Criminal proceedings, 86, 88

Cross-examination, 99, 107, 160:
  interview, 99, 107;
  trial, 160

Crossing arms (interview), 105

Custody of evidence, 31, 39–45, 50–53:
  collecting evidence, 31, 39–40, 43–45, 50–52;
  evidence rules, 41–45;
  evidence tools, 53

Customer at ease (interview), 100

Customer confrontation, 108

Customer contact, 32, 86–88

Customer fraud categories, 59–61:
  residential fraud problems, 59;
  commercial fraud problems, 59–60;
  false claims, 60–61;
  mail fraud, 61

Customer identification, 56, 68–69, 77:
  national identification number, 56, 77;
  name, 77;
  date of birth, 77;
  social security number, 77

Customer information service (CIS), 21, 23

Customer notification, 32

Customer response, 39–40

Customer service, 21–22

Customers amnesty period, 25

Cut seals (procedures), 184–185

# D

Damaged meter (procedures), 185–186

Damaged/removed lock/locking band (procedures), 186

Danger to life, limb, or property (procedures), 187

Dangerous accounts (investigation), 113–128:
  the problem, 113–115;
  drug manufacturers, 115–120;
  indoor marijuana growing, 120–123;
  defusing anger and aggression, 123–126;
  summary, 126;
  recommended actions, 127;
  references, 128

Deception detection (investigative interview), 102–107:
  behavior, 102–107;
  pronouns, 106;
  nouns, 106;
  verbs, 106;
  repeated accounts, 107

Deceptive advertising (fraud scheme), 166

Decision factors (liaison), 149

Defective meters, 37

Definitional lie, 102

Defusing anger and aggression, 123–126

Degree-day analysis, 85

Demand meters not consuming, 37

Denial (overcoming), 108

Denial of service (DOS), 171

Departments (revenue protection program), 21–23

Deposition (pretrial statement), 159

Deregulation challenges, 165–167

Detecting deception (interview), 102–107:
  behavior, 102–107;
  pronouns, 106;
  nouns, 106;
  verbs, 106;
  repeated accounts, 107

Detecting theft of service, 15, 17–19, 26, 29, 35–38, 52:
  revenue protection program, 17–19;
  reward program, 26, 35–36, 52

Detection component (revenue protection program), 17–19

Detection reward program, 26, 35–36, 52

# Index

Developing policies (revenue protection program), 33
Diagram/sketch (crime scene), 47, 157–158
Digital photography, 51:
standard operating procedure, 51;
image preservation, 51;
image storage, 51;
image file save, 51
Direct evidence, 42
Direct examination (trial), 160
Directory of Corporate Affiliations, 73
Disconnect fee, 81
Disconnect service, 31, 81–82, 88, 113:
fee, 81
Disengagement (anger), 126
Dishonest thieves, 9–10
Dispatcher, 184–187
Diversion (electricity), 2
DNA samples, 43
Documentary evidence, 43
Documenting crime scene, 39–40, 45–52, 157:
processing, 39–40, 45–50
Draft city ordinance, 189–192
Drug Enforcement Agency (DEA), 123
Drug manufacturers, 114–120, 127
Drug task force, 150
Dun & Bradstreet (dnb.com), 79
Duties (investigator), 178

# E

Economic perspective, 6–7
Education/experience (investigator), 178
Educational period, 25
Elaborating (interview), 104
Electric Power Act of 1997 (Kenya), 2
Electric/electricity theft (tip information), 197
Electrocution, 3

Electronic meters, 167–168, 175
Emotion (anger), 124
Employee intelligence source, 134–135
Employee theft, 129–132, 146:
attitudes/motivation, 129–130;
investigation, 130–131
Employee training, 18–19, 25–26, 33–34, 39–40, 46–47, 52–53, 110, 127, 151–152
Employees amnesty period, 24
Energy services provider (ESP), 166
Energy theft, x, 1–16, 197:
tip information, 197
England, 5
Escape route plan, 97
Essential elements of information (EEI), 173
Europe, 114
Evidence collecting, 31, 39–40, 43–45, 50–52:
bags, 44
Evidence custody, 31, 39–45, 50–53:
collecting evidence, 31, 39–40, 43–45, 50–52;
evidence rules, 41–45;
evidence tools, 53
Evidence/exhibits (trial), 161
Evidence identification, 44
Evidence rules, 41–45:
real evidence, 42;
circumstantial evidence, 42;
opinion evidence (expert testimony), 43;
documentary evidence, 43;
scientific evidence, 43
Evidence tools, 44, 53
Evolving challenges, 165–176:
impact of deregulation, 165–167;
changes in metering, 167–171;
Sarbanes-Oxley Act of 2002, 171–172;
protecting investigators and investigations, 172–175;
summary, 175;
recommended actions, 175–176;
references, 176

Exceptions report, 173
Excuses (interview), 103
Expert testimony, 43
Expert witness, 43, 158–159
Extension cords (procedures), 185
Eyewitness testimony, 42, 158

## F

Fair and Accurate Credit Transaction Act of 2003, 132
Fair Credit Reporting Act (FCRA), 132
False claims (fraud), 60–61
Falsifying lie, 102
Field interview (investigative), 86, 88, 97
Field notes, 31
Fiji, 114
Final survey (crime scene), 157
Financial pressure (fraud), 57
Fingerprints, 43
Fire investigators/fire department, 149–152, 162–163
Fixer/meter rigger services, 2, 12–13, 95–96, 100, 110, 135–145, 168
Fleeing position (interview), 105
Florida, 115
Foreign Index to the Directory of Corporate Affiliations, 73
Forged endorsement, 63
Forged signature, 62–63
Fraud (energy theft), x, 7–12, 29, 37, 55–69, 76, 183–184:
 categories, 8–12;
 investigation, 55–69, 76
Fraud crime elements, 56–57
Fraud investigation, 55–69, 76:
 fraud problem, 55–58;
 crime elements, 56–57;
 categories of customer fraud, 59–61;
 model, 61;
 prevention, 61;
 detection, 61;
 combating customer fraud, 62–66;
 getting below tip of iceberg, 67, 76;
 summary, 68;
 recommended actions, 68–69;
 references, 69
Fraud model, 61
Fraud problem, 55–58
Fraud schemes targeting consumers, 166–167
Fraud triangle, 57
Fraudulent application for service, 59:
 residential, 59;
 commercial, 59
Fraudulent reconnect, 7–8, 29, 37, 183–184:
 procedures, 184, 186–187
Fraudulent service procedures, 186–187
Freebase (crack/freebase cocaine), 113, 117
Full body motions (interview), 104
Funnel approach (interview), 94, 107:
 model, 107

## G

General questioning stage (interview), 99
Georgia, 114
GHB (rape drug), 116
Government sources (information), 71–72
Green power fraud scheme, 166
Green seal account, 27
Grow house (indoor marijuana growing), 5, 12–14, 114–115, 120–123, 127, 155–156

Guam emergency legislation, 3, 25, 193–195
Guam Power Authority, 26
Guesstimate back bill, 82

## H

Hacker/hacking (computer system), 168
Hands over mouth (interview), 104

# Index

Hidden fees (fraud scheme), 166

High intensity discharge (HID) lighting, 122–123

Honest thieves, 8–9

Human factors, 40–41

Hydro meters, 13

## I

Illegal business theft, 12–13

Illegal reconnect/connection, 7–8, 29, 37, 183–184:
procedures, 184

Illustrations (points), 104

Inaccessible meters, 37

Inactive account, 28–29, 183:
meter reading, 28–29

India, 3–4

Indoor marijuana growing, 5, 12–14, 114–115, 120–123, 127, 155–156

Industrial and commercial cases (information sources), 10–11, 76–77

Inflation protection (fraud scheme), 166

Informant (theft), 134

Information acquisition strategy, 95

Information examination (interview), 99

Information gathered, 95

Information protection (investigation), 172–175

Information sources (revenue recovery), 71–80, 89:
resources, 71–80, 89;
government sources, 71–72;
business sources, 71, 73;
legal and safety sources, 71, 73–75;
internet as investigative tool, 71, 74, 78–80;
theft of service investigations, 76;
fraud investigations, 76;
commercial and industrial cases, 76–77

Information to be acquired, 95

Informational questions, 94, 99–100

Insider information (fraud scheme), 166

Insufficient funds check, 63–64, 68

Intelius.com, 78

Intelligence development activities/tools, 132–135:
trend analysis, 132–133;
sources, 133–135;
surveillance, 133

Intelligence sources, 133–135:
surveillance, 133;
investigation reports/links, 134;
newspapers/other media, 134;
informants, 134;
utility employees, 134–135;
law enforcement officers, 135;
computer reports, 135

Internal investigations, 129–132:
employee theft, 129–132

International Utility Revenue Protection Association (IURPA), 7, 33

Internet as investigative tool (information sources), 71, 74, 78–80, 89

Interrogation technique, 107–109

Interrogator characteristics, 93, 107–109:
factors, 93;
technique, 107–109;
persistence, 109

Interview (investigative), 91–111:
planning, 91–96;
procedures, 92, 97–102, 107–109;
dynamics, 92–93, 97–98;
contamination, 93–94;
effectiveness, 94;
purpose, 95;
constraints, 95–96;
ethics, 96;
scheduling, 96;
field interviews, 97;
recording, 97–99;
office interviews, 98;
stages, 98–99;
detecting deception, 102–107;
advanced techniques, 107–109;
skills practice, 110;
summary, 110;
recommended actions, 110;
references, 111

Interview procedures, 92, 97–102, 107–109:
  guide, 92;
  field interviews, 97;
  office interviews, 98;
  advanced techniques, 107–109

Interview skills practice, 110

Interview/environment comments (customer), 103

Interview stages, 98–99:
  general questioning, 99;
  examination of information, 99;
  cross-examination, 99

Investigation at potentially dangerous accounts, 113–128:
  the problem, 113–115;
  drug manufacturers, 115–120;
  indoor marijuana growing, 120–123;
  defusing anger and aggression, 123–126;
  summary, 126;
  recommended actions, 127;
  references, 128

Investigation component (revenue protection program), 17, 20–21

Investigation follow-up, 40

Investigation initiation, 50

Investigation of crime scene, 35–53, 155–157:
  detecting theft, 35–38;
  principles of investigation, 39–41;
  follow-up, 40;
  rules of evidence, 41–45;
  processing crime scene, 45–49, 155–157;
  documenting crime scene, 49–52;
  initiation, 50;
  summary, 52;
  recommended actions, 52–53;
  references, 53

Investigation of fraud, 55–69:
  fraud problem, 55–58;
  categories of customer fraud, 59–61;
  combating utility customer fraud, 62–66;
  getting below tip of iceberg, 67;
  summary, 68;
  recommended actions, 68–69;
  references, 69

Investigation principles (theft of service), 39–41

Investigation report, 32, 49–50, 134, 199–200:
  reports/links, 134;
  report form, 199–200

Investigative challenges and tools, 129–147:
  internal investigations, 129–132;
  criminal intelligence, 132–135;
  the scenario, 135–138;
  challenge and tools, 138–145;
  summary, 146;
  recommended actions, 146;
  references, 147

Investigative flowchart, 139–140, 145

Investigative interview, 91–111:
  planning, 91–96;
  procedures, 92, 97–102, 107–109;
  dynamics, 92–93, 97–98;
  contamination, 93–94;
  effectiveness, 94;
  purpose, 95;
  constraints, 95–96;
  ethics, 96;
  scheduling, 96;
  field interviews, 97;
  recording, 97–99;
  office interviews, 98;
  stages, 98–99;
  detecting deception, 102–107;
  advanced techniques, 107–109;
  skills practice, 110;
  summary, 110;
  recommended actions, 110;
  references, 111

Investigative matrix, 139, 142, 145

Investigator checklist, 53

Investigator duties, 33–34

Investigator/investigation protection, 172–175

Investigator qualifications/selection, 39

Investigator role, 36

Investigator training, 52–53, 110

Invoking the rule (trial), 160

Israel, 3

Issues/elements coverage, 100–101

## J

Jamaica, 5, 28
Jumper cable, 1, 8, 183
Jury deliberation (trial), 160–161

## K

Ketamine (ket/ketamine hydrochloride), 116–117
Kevdb.infospace.com, 80
Knowledge of sources (information), 71–80:
  government sources, 71–72;
  business sources, 71, 73;
  legal and safety sources, 71, 73–75;
  theft of service investigations, 76;
  fraud investigations, 76;
  commercial and industrial cases, 76–77;
  internet as investigative tool, 71, 78–80
Knowledge/skill (investigator), 178
Ku Klux Klan, 114

## L

Latin America, 5
Law enforcement officers/police department, 127, 135, 149–153, 162–163:
  liaisons, 149–153
Law enforcement/court trial, 149–163:
  establishing liaisons, 149–153;
  solvability factors, 153–155;
  major crime scene investigations, 155–157;
  preparing for/going to trial, 158–162;
  summary, 162–163;
  recommended actions, 163;
  references, 163
Law of torts, 73
Legal sources (information), 71, 73–75
Legislation (Territory of Guam), 193–195
Legitimate business theft, 12
Letter and bill approach, 86–87
Leveler mode phrasing (questions), 94

LexisNexis.com, 79
Liability, 40, 83
Liaisons (law enforcement/court), 149–153
Licensing/certification (investigator), 178
Lie detection (deception), 102–107
Link analysis, 139, 142–145
Load analysis, 85
Load factors information, 110
Local government records, 75
Lock and record method, 84–85
Lost property traceable, 154
Lost revenues collection (revenue recovery), 86–88

## M

Mail fraud, 60–61
Major crime scene (investigation), 155–157:
  basic premise, 156;
  preparation, 156;
  approach, 156;
  secure and protect, 156–157;
  preliminary survey, 157;
  physical evidence, 157;
  narrative report, 157;
  photographs, 157;
  sketches, 157;
  final survey, 157
Management responsibility, 17, 21, 61
Manipulators, 104
Marijuana growing, 5, 12–15, 114–115, 120–123, 127, 155–156
MDMA (Ecstasy), 115–116
Meter bypass, 1–2, 13, 183
Meter control, 18, 26–27
Meter cycling/testing, 27
Meter damaged, 183
Meter data managing agent (MDMA), 170
Meter department, 21–22

Meter diversion, 2

Meter fixer/rigger services, 2, 12–13, 95–96, 100, 110, 135–145, 168

Meter inspection (crime scene), 48–49

Meter maintenance, 7

Meter readers/service personnel, 15, 21–23, 37–38, 177, 184–187

Meter seal/meter control program, 17–18, 26–27, 173

Meter tampering, 2, 12–13, 19, 28, 39, 81, 108, 110, 112–115, 135–145, 168

Meter testing and cycling, 27, 81, 85:
test fee, 81

Metering changes (challenges), 167–171

Methamphetamine/meth labs, 13–14, 114–120:
meth rage violence, 14

Method of operation, 154

Missouri, 116

Models (revenue protection program), 17–21

Moody's International Company Data, 73

Moral excuse proposal, 108

Motive and opportunity (suspect), 155

# N

Name fraud, 183

Name game, 55–56, 67, 69

Narrative description (crime scene), 47, 157

National Bureau of Standards, 74

National Change of Address (NCOA) system, 66

National Fire Protection Association, 74

National identification number, 56, 77

National Labor Relations Board (NLRB), 131

National Reporter System reports, 73

National Safety Council, 74

National Technical Information Service, 74

Ndcr.com, 79

Netherlands, 116

New York, 115

New Zealand, 3, 114

Newspapers/other media, 134

Non-scheduled interview, 91

Non-standardized interview, 91–92

Non-sufficient funds check, 63–64, 68

Nonverbal communication, 93–94, 104–105:
clues of deception, 104–105

Notify customer, 32

Nouns (detecting deception), 106

# O

Oaths (interview), 103

Objections (trial), 160

Office interview (investigative), 86, 88, 98

Open-ended questions, 94, 99

Operations security (OPSEC), 172–176:
operations, 172–176;
physical, 172–176;
personal, 172–176

Operation York Connection, 121, 127

Opinion evidence, 43, 158–159

Opportunity (fraud), 57

Organized crime, 4, 12–14

Organizing program (revenue protection), 17–34:
models, 17–21;
responsibilities, departments, and coordination, 21–23;
proactive activities, 24–29;
reactive activities, 29–32;
developing policies, 33;
summary, 33;
recommended actions, 34

Out of business/back in business (fraud), 60, 65:
combating, 65

# Index

## P

Pakistan, 4, 28

Paper chase (revenue recovery), 71–89:
sources of information, 71–80;
computing back bill, 80–86;
collecting lost revenues, 86–88;
summary, 88–89;
recommended actions, 89

Paralanguage, 94

Patience characteristic (interrogator), 109

Pending tasks, 50

PeopleData.com, 79–80

Peoplefind.com, 79

Peoplefinders.com, 79

Persistence (interrogator), 109

Personal security (investigation), 172–176

Phone.people.yahoo.com, 80

Photography/photographs, 30, 39–40, 43, 46–53, 157–158, 161:
initial, 30;
digital, 51

Physical evidence, 42, 47–48, 155, 157

Physical security (investigation), 172–176

Planned bankruptcy (fraud), 60, 64–65:
combating, 64–65

Planning (investigative interview), 91–96

Police department/law enforcement officers, 127, 135, 149–153, 162–163:
liaisons, 149–153

Policy development, 33–34

Policy review, 15, 68

Position description (field personnel), 52

Position description (utilities service investigator), 177–179:
scope of responsibility, 178;
duties and responsibilities, 178;
education and experience, 178;
knowledge and skill, 178;
special licensing and certification, 178;
working conditions, 179;
reference, 179

Potentially dangerous accounts (investigation), 113–128:
the problem, 113–115;
drug manufacturers, 115–120;
indoor marijuana growing, 120–123;
defusing anger and aggression, 123–126;
summary, 126;
recommended actions, 127;
references, 128

Practice (interview skills), 110

Preliminary survey (crime scene), 157

Prepared questions, 101

Preparing for/going to trial, 158–162

Pressure to commit fraud, 57

Prevention component (revenue protection program), 17–19

Principles of investigation (theft of service), 39–41

Proactive activities (revenue protection program), 24–29, 34:
develop program, 24;
amnesty period for employees, 24;
amnesty period for customers, 25;
training for employees, 25–26;
public awareness campaign, 26;
meter seal and meter control program, 26–27;
testing and cycling of meters, 27;
anti-tampering devices, 28;
strike force, 28;
reading meters at inactive accounts, 28

Problem solving (anger), 124

Procedures (crime scene search), 47–49

Procedures (handling utility theft/fraud situations), 33–34, 53, 181–187:
development, 33–34, 53;
common violations, 183–184;
illegal reconnect, 184;
cut seals, 184–185;
extension cords, 185;
damaged meter, 185–186;
damaged/removed lock/locking band, 186;
yellow seals, 186;
fraudulent service, 186–187;
danger to life, limb, or property, 187;
best judgment, 187

Procedures (investigative interview), 97–102, 107–109;
    field interviews, 97;
    office interviews, 98;
    advanced techniques, 107–109

Procedures development, 33–34, 53

Processing crime scene, 39–40, 45–50

Program development (proactive activities), 24

Pronouns (detecting deception), 106

Property access, 46

Protecting investigators/investigations (challenges), 172–175

Protocols (crime scene search), 45–46

Public awareness, 18–19, 26:
    awareness campaign, 26

Public housing, 67

Public Law #24–31 (Territory of Guam), 193–195

Public Record Finder.com, 78

Pyramid/Ponzi fraud scheme, 166–167

## Q

Question answered with question (interview), 103

Question objective, 93

Question repeats (interview), 103

## R

Rationalization (fraud), 57–58

Reactive activities (revenue protection program), 29–33, 34:
    possible theft detected, 29;
    crime scene secured, 30;
    initial photographs, 30;
    examine scene, 30;
    collect evidence, 31;
    complete field notes, 31;
    disconnect service, 31;
    notify customer, 32;
    back at office, 32;
    contact customer, 32

Real evidence, 42

Recommended actions, 15, 34, 52–53, 68–69, 89, 110, 127, 146, 163, 175–176:
    revenue protection, 15;
    program organizing, 34;
    crime scene investigation, 52–53;
    fraud investigation, 68–69;
    paper chase and revenue recovery, 89;
    investigative interview, 110;
    investigation at potentially dangerous accounts, 127;
    investigative challenges and tools, 146;
    working with law enforcement/ going to court, 163;
    evolving challenges, 175–176

Reconnect fee, 81

Red seal account, 27

Reid technique (interrogation), 107–109

Reinforce sincerity (interview), 108

Reliability test (information), 96

Repeated accounts (detecting deception), 107:
    comparison of events, 107

Report writing, 32

Residential accounts, 10, 14, 37–38, 59, 62, 68, 88:
    fraud problems, 10, 14, 59, 68

Residential fraud problems, 10, 14, 59, 68:
    fraudulent application for service, 59;
    theft of service, 59;
    skips, 59;
    bad checks, 59

Respect (defusing anger), 125

Respectful remarks overuse (interview), 104

Respondent factors, 93

Respondent withholding information, 101–102:
    risk, 101–102

Responsibility, 17, 20–23, 33–34, 177–178:
    revenue protection program, 21–23;
    investigator, 177–178

# Index

Revenue protection (challenge), 1–16:
case studies, 1–6;
economic perspective, 6–7;
risk management issue, 7–8;
theft and fraud categories, 8–12;
organized crime, 12–14;
summary, 14;
recommended actions, 15;
references, 16

Revenue protection models, 17–21, 61

Revenue protection program (organizing), 17–34, 61:
models, 17–21, 61;
responsibilities, departments, and coordination, 21–23;
proactive activities, 24–29;
reactive activities, 29–32;
developing policies, 33;
summary, 33;
recommended actions, 34

Revenue protection/recovery program, x, 1–34, 49–50, 61, 71–89, 91–111, 134, 175, 197, 199–200:
challenge, 1–16;
organizing, 17–34, 61;
models, 17–21, 61;
investigation report, 32, 49–50, 134, 199–200;
paper chase, 71–89;
revenue recovery interview, 91–111;
web page, 197

Revenue protection web page (sample), 197

Revenue recovery paper chase, 71–89:
sources of information, 71–80;
computing back bill, 80–86;
collecting lost revenues, 86–88;
summary, 88–89;
recommended actions, 89

Revenue recovery priority, 61, 95–96, 107, 109

Revolving door game, 55–56, 67, 69

Rewards program (theft detection), 26, 35–36, 52

Risk management, 7–8

Rules of evidence, 41–45:
direct evidence, 42;
real evidence, 42;
circumstantial evidence, 42;
opinion evidence (expert testimony), 43;
documentary evidence, 43;
scientific evidence, 43

## S

Safety inspection, 48

Safety sources (information), 71, 73–75

Sarbanes-Oxley Act of 2002 (challenges), 171–172

Scheduled interview, 91

Scientific evidence, 43

Seal damaged/removed (meter), 37, 81, 183–185:
procedures, 184–185

Seal program, 17–18, 26–27, 173

Search warrant, 46

Security (crime scene investigation), 21, 23, 30, 47, 156–157, 172–176:
operations, 172–176;
physical, 172–176;
personal, 172–176

Selective memory (interview), 103

Service disconnection, 31, 81–82, 88, 113

Service personnel/meter readers, 15, 21–23, 37–38, 177, 184–187

Shocking (fraud scheme), 166

Sketch/diagram, 47, 157–158

Skips (fraud), 59–60, 66, 68–69, 80:
residential, 59;
commercial, 60;
combating, 66, 69, 80

Skip trace program, 69

Skip tracing, 66, 69, 80:
skip trace program, 69

Smartpages.com, 80

Social network analysis, 139–141, 145

Social security number, 77

Software (investigative), 145

Solvability factors (law enforcement/court), 153–155

Sources of information (revenue recovery), 71–80, 89:
resources, 71–80, 89;
government sources, 71–72;
business sources, 71, 73;
legal and safety sources, 71, 73–75;
internet as investigative tool, 71, 74, 78–80;
theft of service investigations, 76;
fraud investigations, 76;
commercial and industrial cases, 76–77

South Africa, 2–3

South America, 117

Speech pattern changes (interview), 102–103

Stages (crime scene search), 47–48

Stalling for time (interview), 103–104

Standard & Poor's Register of Corporations, Directors and Executives, 73

Standard operating procedure (SOP), 51

Standardized interview, 91–92:
scheduled, 91;
non-scheduled, 91

Statement analysis (interview), 105–107:
pronouns, 106;
nouns, 106;
verbs, 106

Stolen meter, 183

Strike force, 28

Supposition for crime, 108

Suspect identification/information, 154–155

Switchboard.com, 80

## T

Taking the stand (trial), 160

Tampering (meter), 2, 12–13, 19, 28, 39, 81, 108, 110, 112–115, 135–145, 168

Tape recording (interview), 97–99

Task force approach, 20–23

Tax assessor files, 75

Telephone contact, 86, 88

Testifying (trial), 158–163

Theft and fraud categories, 8–12

Theft of service (crime scene investigation), 15, 17–19, 29, 35–53, 59–60, 68, 76:
detecting, 15, 17–19, 29, 35–38, 76;
principles of investigation, 39–41;
rules of evidence, 41–45;
processing crime scene, 45–50, 59–60;
documenting crime scene, 45, 49–52;
summary, 52;
recommended actions, 52–53;
references, 53

Theft of service, 10–11, 14–15, 17–19, 29, 35–53, 59–60, 68, 76–77:
commercial, 10–11, 59–60, 76–77;
residential, 10, 14, 59, 68;
detection, 15, 17–19, 29, 35–38, 76;
crime scene investigation, 15, 17–19, 29, 35–53, 59–60, 76;
information sources, 76

Thomas Regional Industrial Directory, 73

Training for employees, 18–19, 25–26, 33–34, 39–40, 46–47, 52–53, 110, 127, 151–152

Trend analysis, 83–84, 89

Trial (court), 149–163:
establishing liaisons, 149–153;
solvability factors, 153–155;
major crime scene investigation, 155–157;
preparing for/going to trial, 158–162;
summary, 162–163;
recommended actions, 163;
references, 163

## U

U.S. Government Printing Office, 72

U.S. Postal Service (USPS), 61, 66

Unauthorized meter, 183

Underwriters' Laboratories, 74

Uniform commercial code filings, 75

United States, x, 1–2, 6, 13, 39, 47–49, 82, 114–117

# Index

Unlisted meters, 37

Utilities service investigator (position description), 177–179:
scope of responsibility, 178;
duties and responsibilities, 178;
education and experience, 178;
knowledge and skill, 178;
special licensing and certification, 178;
working conditions, 179;
reference, 179

Utility communications, 15, 67, 153, 165–167

Utility customer fraud, 62–68:
application for service, 62;
bad checks, 62–64;
planned bankruptcy, 64–65;
out of business/back in business, 65;
skips, 66

Utility distribution company (UDC), 166

Utility employee (intelligence source), 134–135

Utility fraud survey, 58

Utility theft and fraud situations (handling procedures), 181–187:
common violations, 183–184;
illegal reconnect, 184;
cut seals, 184–185;
extension cords, 185;
damaged meter, 185–186;
damaged/removed lock/locking band, 186;
yellow seals, 186;
fraudulent service, 186–187;
danger to life, limb, or property, 187;
best judgment, 187

Utility theft/fraud, x, 2, 8–12, 181–187:
categories, 8–12;
handling procedures, 181–187

Utility theft ordinance (draft), 189–192

# V

Validity test (information), 96

Verbal communication, 93–94, 102–104:
clues of deception, 102–104

Verbs (detecting deception), 106

Verification questions, 94

Videotape recording, 51–52:
evidence collecting, 51–52

Visual inspection (meter), 171, 175

# W–X

Warning indicators (drug manufacturers/marijuana growing), 113–128

Weaker denials given (interview), 104

Web page sample (revenue protection), 197

Whowhere.com, 80

Willingness (interview goal), 100–101

Witness checklist (trial), 162

Witness to crime, 154

Witness (trial), 158, 162:
checklist, 162

Working conditions (investigator), 179

Working with law enforcement/going to court, 149–163:
establishing liaisons, 149–153;
solvability factors, 153–155;
major crime scene investigations, 155–157;
preparing for/going to trial, 158–162;
summary, 162–163;
recommended actions, 163;
references, 163

# Y–Z

Yellow seal account, 27, 186:
procedures, 186

Yellow seals (procedures), 186

Yourownprivateeye.com, 78